Collins *gem*

Birds

Jim Flegg

HarperCollins*Publishers*
77-85 Fulham Palace Road, London W6 8JB

www.collins.co.uk

Collins is a registered trademark of HarperCollins Publishers Ltd.

First published 2001. This edition published 2004.

07

10 9 8 7 6 5 4

ISBN 10 0-00-717860-3

ISBN 13 978-0-00-717860-5

Design by Penny Dawes
Printed in Italy by Amadeus S.r.l.

Contents key

The birds in this book fall into 27 broad family groupings, each of which is identified by a typical silhouette at the top of the page. Birds are highly mobile and versatile creatures, and occasionally their anatomical adaptations to a particular way of life may outweigh family similarities. For example, swallows and swifts, so similar in appearance, are unrelated, and cranes (which look like herons) are actually related to the crakes and rails. In such cases, for ease of use the page heading silhouette is that which makes identification easiest for the birdwatcher.

Divers (Gaviiformes) (pp. 18–19) and Grebes (Podicipediformes) (pp. 20–21) are specialist diving birds of both fresh and salt waters, hunting fish and other small aquatic animals. Divers are slim and short-necked, grebes plumper: both have feet with lobed toes set back near a stumpy tail. Their wings are small, beat rapidly, and they fly relatively infrequently.

Fulmars and Shearwaters (Procellariiformes) (pp. 22–23) are oceanic seabirds with beaks showing clear signs of segmentation and with conspicuous paired tubular nostrils on the ridge. They are masters of energy-efficient gliding, and come ashore only to breed. They feed on fish and plankton caught near the surface.

Gannets and Cormorants (Pelecaniformes) (pp 24–26) are large fish-eating waterbirds, characterised by powerful beaks. Gannets are maritime, spend most time in the air and dive spectacularly; cormorants spend more time on the water (fresh or salt), dive from the surface, and pursue prey underwater propelled by their large feet with all four toes joined by webbing.

Herons and allies (Ciconiiformes) *(pp. 27–32)* **and Crane (Gruiformes)** *(p. 82)* are notably long-legged, long-necked wetland birds feeding on various small animals. Most are large, some huge. They have long, dagger-like beaks and stab at their prey. Some have adapted to drier habitats.

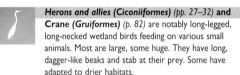

Swans, Geese and Ducks (Anseriformes) *(pp. 33–58)* form a uniform family, generally aquatic (fresh and salt waters), some carnivorous, others vegetarian. Many dabble for food, while others dive. All are characterised by 'duck-like' beaks and by their triangular webbed feet.

Birds of Prey (Accipitriformes and Falconiformes) *(pp. 59–72)* are often called 'raptors', and are characterised by relatively large effective eyes, markedly hooked beaks for tearing flesh (all are carnivorous or scavengers) and by long, usually bare lower legs ending in sharply hooked talons. Females are often substantially larger than males.

Game Birds (Galliformes) *(pp. 73–79)* **and Rails (Gruiformes)** *(pp. 81–84)* are generally omnivorous and characterised by bulky bodies and comparatively small heads. Short rounded wings lift them rapidly into flight. Game birds are terrestrial, with upright stance, running powerfully. Rails are marshland birds, with long legs and large feet. Some swim well, some dive.

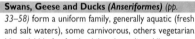

Waders (Charadriiformes) *(pp. 85–116)* are shoreline or marshland birds, feeding on a variety of small invertebrate animals. Most are relatively long-legged, with long toes. Most important identification features to observe (beside plumage colour) are beak length and shape, wing and tail flight patterns and leg colour.

Skuas, Gulls and Terns (Charadriiformes) *(pp. 117–130)* are long-winged web-footed seabirds. Skuas are oceanic or coastal, piratical or predatory, but can fish for themselves. Gulls are more omnivorous, larger species predatory, ancestrally coastal but now often occur inland. Terns are smaller, slimmer, shorter-legged, with longer slimmer wings, and dive from the air for small fish prey. Gulls often, skuas and terns rarely, rest on the water.

Auks (Charadriiformes) *(pp. 131–133)* are robustly dumpy, short-necked seabirds with short narrow wings and whirring flight. They feed on fish caught by diving from the surface and pursue their prey underwater, propelled by their wings.

Pigeons (Columbiformes) *(pp. 135–138)* are heavy-bodied, small-headed vegetarian (largely seed-eating) birds with fast direct flight. There is no distinction between pigeons and doves.

Cuckoo (Cuculiformes) *(p. 134)* and **Nightjar (Caprimulgiformes)** *(p. 146)* are long-tailed birds with short pointed wings. Both are short-legged and have small beaks, feeding on insects, which are caught in flight by nightjars.

Owls (Strigiformes) *(pp. 139–145)* are character-istically stocky, with short tails and an upright stance. Large heads and big eyes surrounded by a prominent facial disc indicate largely nocturnal life-styles. Small animal prey is captured in powerful sharp talons.

Kingfisher, Bee-eater, Hoopoe *(Coraciiformes)* *(pp. 152–154)* form a group with little in common anatomically, but all sufficiently brightly coloured to be readily identified. All are carnivorous, their prey ranging from insects to lizards and fish.

Woodpeckers (Piciformes) *(pp. 155–159)* form a close-knit group, featuring a strong straight dagger-like beak, long, strong central tail feathers used as a prop when perched on trunks, and powerful feet with toes distinctively arranged two pointing forward, two back. Their flight is undulating, their calls strident, their drumming far-carrying.

Larks, Pipits and Wagtails *(Passeriformes: Alaudidae, Motacillidae)* *(pp. 160–169)* are largely terrestrial, swift running birds of open habitats. Larks and pipits are heavily streaked and well camouflaged, with a long hind claw; wagtails are more colourful, with long incessantly wagged tails. All eat insects and small soil invertebrates, larks also eat vegetable matter.

Swallows and Martins *(Passeriformes: Hirundinidae)* *(pp. 149–151)* and **Swifts** *(Apodiformes)* *(pp. 147–148)* have short legs, small beaks, streamlined bodies and slim curved wings. Much time is spent on the wing, including catching insect prey and drinking. Swallows and swifts are taxonomically unrelated, but evolution has shaped the outward anatomy of both groups to suit a common life style.

Wren, Dipper, Dunnock (Passeriformes: Troglodytidae, Cinclidae, Prunellidae)
(pp. 171–1732) a grouping of convenience, rather than indicating close relationship. All, though, are predominantly brown in plumage, largely terrestrial in habit, (the dipper aquatic) and feed on small invertebrate animals, the wren using a finely pointed beak, the others more robust. The sexes are broadly similar.

Thrushes and Chats (Passeriformes: Turdidae)
(pp. 174–189) form an obviously coherent grouping with two major types. Thrushes are larger, stouter-legged and rather longer-tailed, often with a horizontal body posture. Chats are smaller, rounder in the body, with longer, slimmer legs and characteristically flick wings and tail. All share a medium-length pointed beak, stronger in some than others, and have a mixed diet of invertebrate animals augmented by berries.

Warblers and Crests (Passeriformes: Sylviidae)
(pp. 190–209) also form a coherent grouping of small birds, mostly migrants, dividing into distinctive sub-groups, the tiny generally greenish, active canopy-feeding leaf warblers and crests; the brown, sometimes streaked, reedbed Acrocephalus warblers and their allies; and the more robust and colourful Sylvia warblers, where the sexes differ in plumage. All have

shortish insectivorous beaks, and depend heavily on insect food, though turning readily to fruit to augment their diet in autumn.

Flycatchers (Passeriformes: Muscicapidae) *(pp. 210–211)*. Small, migrant, and rather warbler-like, flycatchers share the habit of catching insect prey in flight. They are short-legged, giving an elongated, horizontal perching posture. Their beaks though short and pointed are broad, with bristles round the gape to increase their catching area, and often close with an audible snap.

Tits and Allies (Passeriformes: Paridae, Aegithalidae, Timaliidae) *(pp. 212–219)*. True tits are small, active and agile woodland birds with relatively strong legs and a stubby but powerful beak well suited to an omnivorous diet. They nest in holes which they may either excavate or modify. Long-tailed and bearded tits are not closely related, but possess tit-like beaks and agility. These build complex nests in vegetation.

Nuthatch and Treecreeper (Passeriformes: Sittidae and Certhiidae) *(pp. 220–221)*. Short-legged and with strong feet, these spend much time clinging to trunks and branches. Nuthatches are woodpecker-like in beak and habits (but lack strong central tail feathers). Treecreepers have large eyes and longish, finely pointed beaks to extract insect prey from crevices in the bark.

Shrikes (Passeriformes: Laniidae) *(pp. 224–225)*. A close-knit group of relatively long-tailed, thrush-sized birds, with falcon-like hooked and notched beaks for grasping and tearing small animal and insect prey, which they sometimes impale on thorns for later consumption. Usually favour exposed perches.

Crows *(Passeriformes: Corvidae) (pp. 223, 226–232).*
Large among the Passeriformes, crows are usually gregarious in habit and omnivorous in diet. All have powerful beaks, and are opportunist predators as well as scavengers. The true crows are black or blackish, related genera are more strikingly coloured (eg magpies, jays).

Oriole, Starlings, Waxwing *(Passeriformes: Oriolidae, Sturnidae, Bombycillidae) (pp. 222, 233, 170)* are similar in size and shape, flying fast and straight on triangular wings. Short-medium length straight beaks suit a mixed diet of fruit and invertebrate animals. Plumage and calls are best distinctive features.

Sparrows, Buntings and Finches *(Passeriformes: Passeridae, Emberizidae, Fringillidae) (pp. 234–253)* are small, stocky, rather short-legged, predominantly seed-eating birds, often gregarious. Their beaks are stout and roughly wedge-shaped. Upper mandible ridge is convex in sparrows; lower mandible distinctively larger than upper in buntings. In finches the precise size and shape of the generally triangular beak is a guide to diet and often useful in identification.

Please note the colours in the maps indicate the following: dark grey, resident; mid-grey, winter visitor; light grey, summer visitor.

Introduction

This book gives easily-used identification details linked to
full colour photographs, plus information on habitat, food,
song and behaviour, for over 230 European birds,
covering the species that most birdwatchers could expect
to see in a lifetime. This introduction provides advice on
how birds use the various habitats in our countryside, on
the equipment needed for birdwatching, and on how to
get the best out of your birdwatching by developing all-
important fieldcraft skills.

PARTS OF A BIRD

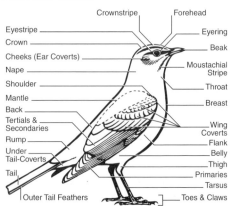

Crownstripe
Forehead
Eyestripe
Eyering
Crown
Beak
Cheeks (Ear Coverts)
Moustachial
Stripe
Nape
Shoulder
Throat
Mantle
Breast
Back
Tertials &
Secondaries
Wing
Coverts
Rump
Flank
Under
Tail-Coverts
Belly
Thigh
Tail
Primaries
Tarsus
Outer Tail Feathers
Toes & Claws

HABITATS AND MIGRATION

Unpredictability is very much a feature of bird life. The most distinctive character of birds (apart from their unique covering of feathers) is their ability to fly. Birds migrate to take advantage of opportunities in one part of the world which last only for part of the year – for example food being plentiful in one area in summer and another in winter. This explains the kaleidoscopic seasonal changes to be seen as migrants depart or arrive. Migration may be over huge distances, from the Arctic Circle to southern Africa or beyond, or be comparatively short-haul. Although this Gem guide covers the birds of the whole of Europe, many birds which do not breed or overwinter in your area will be seen as they come through on passage in spring and autumn. The ability to migrate (occasionally getting blown off course), coupled in many birds with an opportunistic approach to feeding, means that most birds do not necessarily conform to land boundaries laid down by geographers. Migration adds a great deal to the richness of birdwatching.

Nor do birds always stay neatly in the habitat categories that ecologists have attempted to draw for them. Although most of the ducks are associated with the sea or freshwater, kingfishers and dippers with rivers and streams, many birds are not so tidy. Kestrels may be seen over towns, coasts, moorland and motorway, and the gulls are as much at home on farmland and rubbish tip as at sea. Even Blue Tits, traditionally year-round woodland birds, may in midwinter be found in an oakwood, or (equally likely) feeding on an exotic food like peanuts in gardens (or opening milk bottles on town doorsteps to remove the cream), or feeding on insects hibernating in the shelter of the reed stems in a huge, tree-less marsh! In this guide, for each bird we identify the main habitats frequented.

PLUMAGE

Even the plumage of birds changes with time. Feathers are the most obvious external feature of birds, and in many cases are brightly coloured and patterned. Though to us attractive, this patterning serves practical purposes such as attracting a mate, defending a territory, or helping with camouflage. For birdwatchers, plumage is often one of the best identification aids, but there are reasons for caution: as a general rule, males tend to be brighter, and females and immatures duller, as for them camouflage is more important than display. But in many species, the males in winter may also be drab, their bright colours only appearing as the feathers gradually wear down early in the spring.

In most birds, particularly the smaller ones, wear and tear during the year is balanced each autumn by the process called moult, when old feathers gradually fall out and are replaced by new ones. Young birds, two or three months out of the nest, lose their juvenile, often speckled plumage and grow the coloured feathers of the adult for the first time. As this happens, their plumage is a confusing patchwork of old and new. In larger birds like some gulls and birds of prey, the change from juvenile to adult plumage is gradual over three or four years, giving a series of immature plumages making identification quite a problem.

BIRDWATCHING EQUIPMENT

Perhaps the one essential piece of equipment is a pair of binoculars: with these, distant black dots take on an identifiable shape and colour, or the beautiful feather detail of a Great Tit feeding on a garden peanut-holder can be revealed. Binoculars range from cheap to expensive, so a choice over cost and which magnification you will need must be made. There are a few simple guidelines: the binoculars are for you, so they should be comfortable in the hand, easy to use and comfortable hanging round your neck — so do test them outdoors before purchasing. Optical quality tends to increase with price, so with cheaper binoculars (many of which are perfectly satisfactory), check that there are no colour fringes to the images you see, and that telephone poles are not 'bent' by poor lens design. If you wear spectacles, check that you can use the binoculars without removing (or scratching) them. As to magnification, generally avoid more than x10 as they are sensitive to shaking, and let in rather little light. For garden, field and woodland watching, x7 or x8 should be right, if possible with a 'wide angle' field of view. For those who birdwatch mostly on moorland, the coast, estuaries or large reservoirs, x10 is probably the ideal, although these binoculars will be heavier.

A notebook in which you can jot down notes of numbers of species and plumage details and make sketches (particularly of birds new to you) is also a necessity. Take notes immediately you see a new bird: these could turn frustration into satisfaction as the identification problem is later resolved. A pocket fieldguide such as this Gem should always be with you - in your pocket.

As to clothing, common sense is the best guide, but there are points to remember. Avoid bright colours and noisy rustling fabrics. Remember that in exposed habitats, the weather can change (usually for the worse) with surprising speed, so it is always worth having ample warm, wind and waterproof gear. A heavy sweater and a lightweight nylon kagoul or anorak covers most circumstances. On the feet, trainers or baseball boots may be adequate, but on rougher terrain, walking boots may be desirable, and wellingtons are obviously a necessity in wetland habitats. And always take a supply of energy-rich food and drink on longer walks.

FIELDCRAFT

In fieldcraft, the aim is to see birds well without being seen yourself. Experienced birdwatchers rely very much on their ears to give early warning of what is about. Knowledge of bird calls and songs is an invaluable aid to identification: experience is the best teacher, but listening to commercially available recordings is an excellent foundation. Birds also use their ears — the less chatter, laughter and cracking of twigs, the closer you will get. Pause often to listen and look, preferably in a sheltered spot with a good view. Try to merge into the background, using natural features like banks, hedges and sea walls to avoid standing above the skyline. On the coast or in estuaries, check the tide times when planning your visit: often at low tide, the birds will be far out of sight on the mud, at high tide they may have flown off to roost. Remember it is tide times, not night or day, that govern the movements of birds in these habitats. Plan to visit on a rising (preferably) or falling tide for best views, or locate high-tide roosts by watching the flight-lines of waders heading for them — then you could get excellent views. In inland habitats like woodland, again a knowledge of behaviour can help. Birds tend to be active soon after dawn and before dusk, and in summer these are good times to see the singers, and most usefully, become familiar with their songs. In winter, many birds will gather just before dusk to go to roost. In contrast, the middle hours of the day can often be relatively quiet, with few birds moving, especially in a hot summer.

The hides which a feature of many nature reserves often give excellent views and help you to become familiar with the birds of a particular habitat, and to gain an insight into their daily lives. It is easy to forget that a car, strategically parked, can provide similarly good views; or that our homes offer a privileged view of the birds nearby, especially if the garden has some drinking water and plenty of food both on the bird table and in the form of berried shrubs and other food plants. An amazing range of birds can visit even the average suburban garden.

Although fieldcraft is all about getting close to birds without causing disturbance, always remember to put the birds' interests first. Nesting birds not only demand special consideration, but are mostly protected by law, not just from egg collectors but from any disturbance. By all means watch and enjoy garden birds' nests or follow the progress of tit families using nestboxes. A bicycle mirror fixed to a stick allows you a good view of the nest and contents without leaving a tell-tale track through the surrounding vegetation. If the parent is sitting, pass quietly by and come back later. The Birdwatchers' Code also requires that special consideration is given to newly arrived migrants, tired after a long journey, and to all birds during severe winter weather. In both cases it is vital that they can feed uninterrupted, so remember their needs and do not be over-anxious for a good view. And of course, obey the Country Code, leaving gates shut, keeping to footpaths, controlling dogs, causing no fires and leaving no litter. In essence, respect and protect the whole countryside as a valuable asset.

Gavia stellata

ADULT Large (55 cm), slender-bodied diving waterbird, well streamlined for swimming. Watch for slim, pale, upturned beak. Looks hump-backed in flight, with rapid deep wingbeats. Throat often looks blackish. In winter grey, flecked white above, white below. Sexes similar.

JUVENILE As winter adult.

HABITAT Breeds on N European coasts and moorland lakes, winters on coastal seas, also on fresh waters.

NEST Always close to water.

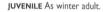

VOICE Cackling calls, breeding season only.

adult
winter

GENERAL Widespread in habitat, never numerous. Black-throated Diver (*G. arctica*): straighter dark beak; Great Northern Diver (*G. immer*): heavy angular head, massive beak.

ADULT Large, stout diver (75cm). Heavy, angular head, large beak (may have a bump on the forehead). Long wings and feet, latter project beyond the tail in flight. Shallow wing beats. Black head and neck with check upperpart in summer; dark grey, with barring in wint Pale eye ring.

adult winter

JUVENILE As winter adult with scaly pattern on upperparts.

HABITAT Winters on Atlantic coasts.

NEST Mound of waterside vegetation.

VOICE Loud wailing calls in breeding season.

GENERAL Does not breed in Britain. Black-throated Diver *(G. arctica)* has grey head in summer, and no eyering in winter.

Podiceps cristatus

adult
winter

ADULT Medium (45 cm), slim, slender-necked diving waterbird. Chestnut and black ruff and crest reduced to black crown in winter. Hump-backed in flight, showing white wing patches. In winter, pale grey back, with black, not yellow, dagger-like beak. Sexes similar.

JUVENILE Chick grey with black stripes; later as winter adult.

HABITAT Widespread except in far N; breeds and winters on large fresh waters; also winters on sea.

NEST Raft of waterweed moored to reeds.

VOICE Guttural croaks and honks in summer.

GENERAL Rarer Red-necked Grebe (*P. grisegena*): smaller, with rufous neck and white cheeks in summer, grey with whitish cheeks in winter.

ADULT Smallest grebe (25 cm). A dark and dumpy diving waterbird, short-necked and tail-less in appearance. Chestnut throat often appears blackish. Watch for short dark beak, pale-tipped in summer. In winter, dull brown above, paler below. Rarely flies far except at night, escapes threats by submerging. Sexes similar.

adult winter

JUVENILE Chick grey with black stripes; later as winter adult.

HABITAT Widespread except far N; breeds and winters on fresh well-vegetated waters; some winter on sheltered coastal seas.

NEST Raft of waterweed, moored to reeds.

VOICE Far-carrying whinnying, usually when breeding.

GENERAL Commonest and most widespread grebe.

Manx Shearwater

FULMARS

Puffinus puffinus

ADULT Medium-sized (35cm). Uniformly black above and white below, No capped effect and no white at the base of the tail. Flies over water tilting from side-to-side. alternating long glides on stiffly-held wings, with shallow flaps.

JUVENILE As adult.

HABITAT Breeds in colonies on islands and cliffs facing the Atlantic. Range extends west and north outside breeding season.

NEST In burrows.

VOICE Loud wails, and cackles.

GENERAL Great Shearwater (*Puffinus gravis*) is larger, and has a black cap and a white patch at the base of the tail.

adult

ADULT Medium seabird (45 cm), superficially gull-like, but actually a petrel. Watch for dumpy but well-streamlined body, dark eyes, and stubby yellow beak with tubular nostrils. Distinctive flight on short, straight all-pale wings, often held slightly downcurved. Glides often, with only occasional wingbeats except near cliffs. Sexes similar.

JUVENILE As adult.

HABITAT Breeds colonially on N and W coasts, usually on cliffs; winters in coastal and oceanic seas.

NEST Single large egg laid on bare ground.

VOICE Cackles and croons on breeding ledges.

GENERAL Widespread, locally common; a successful bird, spreads and colonizes new areas, even on buildings.

adult

Gannet

Sula bassana

juvenile

ADULT Huge (90 cm) unmistakable seabird. Watch for white, cigar-shaped body and long straight, slender, black-tipped wings. In summer, yellow head of adult inconspicuous. Plunges spectacularly for fish. Sexes similar.

JUVENILE Grey-brown, flecked white becoming whiter, reaches adult plumage after three years.

HABITAT Breeds colonially on cliffs on N and W coasts, dispersing to winter at sea.

NEST Mound of seaweed on bare rocky ledge.

VOICE Harsh honks and grating calls at colony.

GENERAL Widespread, but breeding colonies few though sometimes enormous.

ADULT Huge (90 cm), dark, broad-winged seabird. Watch for thick-necked, heavy-beaked appearance; whitish face of breeding adult. Swims well, diving frequently, emerging to dry wings in heraldic stance. Flies straight, often in groups in V formation. Sexes similar.

JUVENILE Dark brown above with paler underside.

HABITAT Breeds colonially on most rocky coasts, occasionally in trees beside large fresh waters. Winters on coastal seas and larger inland waters.

immature

NEST Untidy mound of seaweed and flotsam on rocks.

VOICE Deep grunts at colony.

GENERAL Widespread in most coastal waters, including shallow muddy bays and estuaries avoided by Shags.

Phalacrocorax aristotelis

ADULT Large (75 cm), dark, slim-bodied seabird often with greenish sheen. Watch for slender neck and comparatively slim but hooked beak. Has tufted crest in spring. Lacks white patches of Cormorant (p.22), but has yellow gape. Swims well, diving frequently. Sexes similar.

JUVENILE Dark brown above, unlike Cormorant only slightly paler on underparts.

HABITAT Breeds colonially on rocky coasts. Winters on coastal seas, rarely on inland fresh waters.

NEST Bulky and untidy mound of seaweed, often under rocky overhang.

VOICE Harsh grunts at colony.

immature

GENERAL Widespread, but usually less numerous than Cormorant, favouring deeper and clearer sea.

adult

ADULT Large (75 cm), extremely well-camouflaged, brown heron. Finely streaked and mottled plumage strikingly beautiful at close range. Watch for short, dagger-shaped beak and hunched posture. Despite its size, slips imperceptibly and silently between reed stems: if disturbed, freezes in upright posture. Sexes similar.

JUVENILE As adult, but duller.

HABITAT Year-round resident much of central Europe, summer visitor further N. Favours large freshwater reedbeds.

NEST Reed platform among the reeds.

VOICE Distinctive 'foghorn' booming in breeding season.

GENERAL Inconspicuous, more often heard than seen. Declining in numbers in N and W of range.

Egretta garzetta

ADULT Large (55 cm), slim, all-white heron. Watch for black dagger-like beak, black legs with yellow feet distinctive in flight. Breeding adult has fine filamentous summer plumes on throat and back. Sexes similar.

JUVENILE Duller white, lacking plumes.

HABITAT Summer visitor, breeding colonially in S European wetlands, sometimes resident year-round.

adult

NEST Bulky, in trees overhanging water.

VOICE Honks and shrieks in breeding season.

GENERAL Locally quite common. Rare Great White Egret (*E. alba*): larger, with heavy yellow beak. Scarce Cattle Egret (*Bubulcus ibis*): dumpy, with yellow beak and legs. Squacco Heron (*Ardeola ralloides*): brown-streaked crown and buff back contrasting with white wings. All largely restricted to S Europe.

Grey Heron

Ardea cinerea

ADULT Huge (90 cm), long-legged, long-necked heron. Broad, heavily fingered wings and ponderous flight, legs outstretched but neck folded back between shoulders. Paces slowly through shallow water, stabbing fish and other prey. Yellow dagger-like beak may be orange in summer, when plumes on neck and back conspicuous. Sexes similar.

JUVENILE As adult, but drabber, no crest or plumes.

HABITAT Year-round resident or migrant over much of Europe, summer visitor to N and E. Favours lakes or marshland; in winter also coasts and garden ponds.

NEST Bulky, in trees or reedbeds.

VOICE Amazing cacophony of honks and shrieks at nest, elsewhere typically frank if disturbed.

GENERAL Widespread throughout Europe.

adult

White Stork

Ciconia ciconia

DISTRIBUTION
Summer visitor
to SW and
central/NE
Europe.

ADULT Huge (100 cm) and
superficially heron-like with long
neck and legs and powerful dagger-
like beak, but storks fly neck
outstretched on broad, heavily
fingered wings. Black and dirty-white
plumage: breeding adult has shaggy
throat feathers in summer. Sexes
similar.

JUVENILE As drab adult, brown beak and legs.

HABITAT Feeds fields, marshland, nests in
trees or buildings.

adult

NEST Conspicuously bulky.

VOICE Rarely vocal: grunts and hisses at nest.

GENERAL Beak clattering displays between pairs
at nest more than compensate for lack of true vocalization.
Migrates in flocks over long-
established routes.
Regular in breeding
areas, scarce
elsewhere.

Spoonbill

Platalea leucorodia

DISTRIBUTION
Summer visitor to widely separated major wetlands in SW, north-central, and SE Europe.

ADULT Huge (80 cm), long-legged, long-necked waterbird with distinctive beak and feeding technique as it sifts through mud and shallows in search of molluscs. All-white plumage, legs and beak black. Usually flies in groups, necks outstretched, in V-formation. Sexes similar.

JUVENILE As drab adult, but with black wingtips conspicuous in flight.

HABITAT Feeds in fresh, brackish or saline waters.

NEST Breeds colonially, usually on reed platforms in extensive reedbeds.

adult

VOICE Rarely vocal: grunts at nest.

GENERAL Regular at breeding sites; vagrant elsewhere.

Greater Flamingo

HERONS

Phoenicopterus ruber

DISTRIBUTION
Year-round resident or short-haul migrant to breeding areas in S Spain and France.

ADULT Huge (125 cm), unmistakable waterbird, with very long pink legs and long neck, held outstretched in flight when rich pink wings show to best effect. Usually in flocks, flies in loose V-formation. Distinctive banana-shaped beak is held upside down under water, filtering out food.

JUVENILE Greyer, pink after a year or two.

HABITAT Favours extensive, usually shallow, saline or brackish lagoons with mudflats.

NEST Like miniature volcano, built of mud.

VOICE Goose-like honks and cackles.

GENERAL Locally common near breeding sites, vagrant elsewhere. Other flamingo species regularly escape from waterfowl collections, and may turn up anywhere.

adult

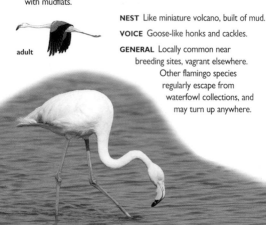

ADULT Huge (150 cm) and familiar. Watch for S-curved neck and orange-red beak with black knob more prominent in male. Arches wings like sails in defence of territory or young. In flight, broad all-white wings produce distinctive creaking sound. Long pattering take-off run. Sexes broadly similar.

JUVENILE As adult, but grey-buff.

HABITAT Widespread across N and W Europe, summer visitor in N, year-round resident elsewhere. Favours fresh waters of most types, including urban areas. Occasional only on sheltered seas.

adult

NEST Bulky mound of reeds etc beside water.

VOICE Often silent, hisses, grunts in breeding territory.

GENERAL Widespread, only locally numerous. Numbers currently recovering after decline due to lead poisoning.

SWANS

Cygnus cygnus

Whooper swan – adult

ADULT Huge (150 cm) all-white swan. Watch for long, straight neck and wedge-shaped head profile. Beak characteristically black and yellow. Wings quiet in flight, but family parties are wonderfully vociferous. Sexes similar.

JUVENILE As adult, but pale grey-buff.

HABITAT Summer visitor breeding on Arctic tundra, migrant or winter visitor to NW Europe, primarily Britain and Ireland. Favours marshland, estuaries and larger fresh waters.

Bewick's swan – adult

NEST Bulky mound of vegetation near water.

VOICE Vocal; bugle-like whooping calls.

GENERAL Scarce, but locally regular. Bewick's swan *(C. bewickii):* much smaller (120 cm), with short, straight neck and shorter black and yellow beak, musical goose-like honking calls.

Pink-footed Goose

Anser brachyrhynchus

ADULT Large (65 cm), neatly-built grey goose. Watch for distinctive dark brown head and short neck, stubby dark beak with pink marking. Legs pink. In flight, grey back and forewings contrast with dark flight feathers. Sexes similar.

JUVENILE Similar to adult.

HABITAT Breeds on tundra, usually on rocky outcrops. Winters across NW Europe on fresh and salt marshland, open farmland, often roosting on large lakes.

NEST Down-lined cup on ground.

VOICE Vocal; distinctive **wink-wink-wink**.

GENERAL Locally numerous. Scarcer Bean Goose (*A. fabalis*): similar, but larger (75 cm), longer necked, with large, dark, wedge-shaped beak with yellow markings, legs orange.

adult

Anser albifrons

ADULT Large (70 cm), grey goose. Watch for adult's white forehead and blackish barring on breast. Beak pink (Russian race) or orange-yellow (Greenland race). In flight shows uniformly grey-brown wings with darker flight feathers. Sexes similar.

JUVENILE Lacks white face and black barring.

adult

HABITAT Breeds on high Arctic tundra, winters on NW and extreme SE European coastal and sometimes inland marshes and open farmland.

NEST Down-lined cup on ground.

VOICE Vocal; high-pitched, musical yelping.

GENERAL Fairly widespread, usually in flocks and locally numerous. Rare Lesser White-fronted Goose (*A. erythropus*): smaller, with more white on head and yellow eye-ring.

adult

ADULT Largest (80 cm) and heaviest-built of the grey geese. Watch for thick neck and dark brown head, pink legs and heavy pink (eastern race) or orange (western) beak. In flight shows pale grey forewing patches. Sexes similar.

JUVENILE As dull adult with brown beak and legs.

HABITAT Summer visitor, year-round resident or winter visitor in NW Europe, winter visitor in S. Breeds on moorland and tundra, winters on farmland and coastal marshes. May roost on freshwater lakes.

NEST Bulky, down-lined cup on ground.

VOICE Vocal; cackling and gabbling calls indicate its ancestry of the farmyard goose.

GENERAL Fairly widespread, frequently introduced by man. Locally numerous.

Branta canadensis

ADULT Largest (75 cm) black goose. Watch for long black neck and head with white chin patch. In flight, uniformly scaly brown wings contrast with black and white rump and tail. Sexes similar.

JUVENILE Resembles dull adult.

HABITAT Occasional vagrants from N America winter with grey geese on marshland. Most European birds (NW and W areas) derive from stock introduced by waterfowl enthusiasts and are year-round residents on larger fresh waters and adjacent grassland.

NEST Bulky down-lined cup, usually near water.

VOICE Strident *aah-honk*.

GENERAL Though restricted to a handful of NW European countries, is increasing in numbers and spreading.

adult

ADULT Large (60 cm), very dark goose. Looks short necked, with stubby beak. Adult has white collar mark. Breast grey in extreme W birds, blackish elsewhere. Watch for all-black neck and wings contrasting with white rump in flight. Sexes similar.

JUVENILE As adult, white bars on back and wings.

adult

HABITAT Breeds on high Arctic tundra, winter on W estuaries, sheltered bays and nearby field

NEST Down-lined cup on ground near sea.

VOICE Vocal; grumbling *rrruk*.

GENERAL Local, but often numerous winter visitor. Barnacle Goose (*B. leucopsis*): similar size, scaly grey back, white belly and white face contrast with black neck. Winter visitor, favours coastal grassland.

ADULT Large (60 cm), rather long-necked goose-like duck with unmistakable black, white and chestnut plumage. Looks pied at a distance and in flight, when chestnut areas are less conspicuous. Sexes similar, but drake has knob on red beak, duck may have white face patch in summer.

JUVENILE Duckling striped black-and-white, immature greyish above, white below.

HABITAT Summer visitor to S Scandinavia, year-round resident in W, winter visitor further S. Favours estuaries and sheltered sandy or muddy bays, occasional inland.

adult

NEST Usually concealed in burrow, deserted building or dense vegetation; down-lined.

VOICE Whistles and barking *ack-ack*.

GENERAL Widespread coastally, often numerous.

adult female

adult male

ADULT Medium (45 cm), surface-feeding duck. Drake handsome, duck subdued in camouflage plumage of distinctive cinnamon browns. In flight, both sexes show green speculum; duck shows distinctive white belly, drake bold white patches on inner half of wing.

JUVENILE Resembles female, as does eclipse male.

HABITAT Summer visitor to N, breeding beside tundra pools. Winter visitor to much of central and S Europe, occurring on lakes, marshes, estuaries and coastal seas.

NEST Well-concealed, down-lined grass cup on ground.

VOICE Drake has characteristic piercing whistle; duck a soft, low purr.

GENERAL Widespread and often common in winter.

Anas strepera

adult female

adult male

ADULT Large (50 cm), surface-feeding duck, apparently drab except at close range, when beautiful detail is apparent. Drake appears overall dull grey, duck well-camouflaged in browns. Watch for black undertail (drake) and distinctive black and white speculum in both sexes in flight.

JUVENILE Resembles female, as does eclipse male.

HABITAT Year-round resident in some central and W areas, summer visitor to N, winter visitor to S Europe. Breeds on marshland beside large fresh waters; winters in similar areas and on estuaries and sheltered coastal waters.

NEST Well-concealed, down-lined grass cup on ground.

VOICE Rare; drake whistles softly, duck quacks.

GENERAL Widespread, rarely numerous, but increasing.

ADULT Medium (35 cm), but distinctively small for a surface-feeding duck. Head

pattern of drake clear only at close range. Duck finely streaked grey-brown. Both sexes have white-bordered dark black and green speculum.

JUVENILE Resembles female, as does eclipse male.

HABITAT Year-round resident over much of Europe, summer migrant in far N and winter visitor in extreme S. Breeds on marshland with pools; winters on well-vegetated fresh waters, and on estuaries and sheltered coasts.

NEST Well-concealed in waterside vegetation.

VOICE Drake has distinctive bell-like call and harsh *krit*; duck a harsh quack.

GENERAL Widespread, often common. Fast and agile, jinking flight is useful guide.

adult female

adult male

Mallard

Anas platyrhynchos

ADULT Large (58 cm), familiar, surface-feeding duck, the drake brightly coloured, the duck well camouflaged in browns and fawns. In flight, both sexes show purple speculum, bordered in white, on trailing edge of inner half of wing.

JUVENILE Resembles female, as does eclipse male.

HABITAT Year-round resident over most of Europe, summer migrant in far N. Seen on all types of fresh waters anywhere; estuaries and coastal seas, especially in winter.

NEST Of grass, lined with dark down, well concealed in ground vegetation, often close to water.

VOICE Drake whistles quietly; duck quacks harshly.

GENERAL Widespread and common, one of the most adaptable of all birds, associates readily with man.

adult
male

adult
female

ADULT Medium (50 cm), surface-feeding duck, swimming low in the water, head tilted down because of massive spoon-shaped beak. Green head of drake looks black at a distance. Duck pale brown with darker speckling. In flight, watch for rapid wingbeats and head-up, tail-down attitude in both sexes, and pale grey forewing patches.

JUVENILE Resembles female, as does eclipse male.

HABITAT Summer visitor to N and E Europe, year-round resident in W, winter visitor to S. Breeds and winters on marshes with shallow muddy lakes, also on reservoirs and sheltered coasts.

adult female

NEST Well-concealed, down-lined grass cup on ground.

VOICE Drake *tuk-tuk*; duck a quiet quack.

GENERAL Widespread and fairly common.

adult male

DUCKS

Anas querquedula

adult
female

adult
male

ADULT Medium (38 cm), surface-feeding duck, only slightly larger than Teal (p.43). Drake striking, watch for bold white eyestripe, duck camouflaged in browns, but has striped face pattern (distinguishes from Teal). In flight, both sexes show distinctive pale blue-grey forewing and white-bordered green speculum.

JUVENILE Resembles female, as does eclipse male.

HABITAT Summer visitor to extensive reedy freshwater wetlands in central and N Europe.

NEST On ground near water, concealed in dense vegetation.

VOICE Drake has distinctive crackling rattle; duck a short quack.

GENERAL Widespread, rarely numerous: unlike Teal, occasional on brackish water and rare on salt waters.

ADULT Medium-sized, surface-feeding duck; drake's distinctively slim, long tail takes the overall length to 70 cm. Watch for white neck mark, prominent even at a distance. Duck pale grey-brown with bold, dark brown markings. Both sexes slim and elongated in flight, with inconspicuous brown speculum on trailing edge of narrow wings.

JUVENILE Resembles female, as does eclipse male.

HABITAT Year-round resident in central and W Europe, summer visitor to N and winter visitor to S. Breeds beside moorland and tundra pools, winters on sheltered coastal waters, estuaries and marshes, occasionally on fresh waters.

NEST Well-concealed, down-lined grass cup on ground.

VOICE Rare; drake whistles, duck growls.

GENERAL Widespread, but rarely very numerous.

adult
female

adult
male

adult
female

adult
male

ADULT Medium (45 cm) diving duck.
Watch for finely-marked grey back of
drake. Duck brown above, shading to white
on belly; rich brown head with large white
face-patch at base of beak. Note golden eyes
and grey beak in both sexes. In flight, both
sexes show bold white wingbar.

JUVENILE Resembles female, as does eclipse male, lacking white
face patch.

HABITAT Summer visitor breeding on N tundra; in winter favours
NW coastal seas, occasionally on fresh waters.

NEST Well-concealed down-lined grass cup on ground.

VOICE Rarely vocal; drake uses low whistle, duck a double quack.

GENERAL Generally scarce, but locally common in winter.

ADULT Medium (42 cm), dumpy, frequently-diving duck. Drake has pied plumage and drooping crest. Duck also compactly built, dark brown above, paler on belly, sometimes with slight crest and small white patch at base of beak. Watch for narrow white wingbar in flight.

JUVENILE Resembles female, as does eclipse male.

HABITAT Summer visitor to N Europe, year-round resident in W, winter visitor in S. Breeds beside reedy lakes and ponds, winters on many types of still, fresh waters.

NEST Well-concealed, down-lined grass cup on ground.

VOICE Rarely vocal; drake uses soft whistle, duck a growl.

GENERAL Widespread, familiar and often common.

adult female

adult male

Melanitta nigra

adult female

adult male

ADULT Medium (50 cm), heavily-built sea duck. Watch for heavy beak, slightly knobbed and black and yellow in summer drake. Unique among wildfowl in its all-black plumage. Duck dark brown with paler buff cheeks. Flies in straggling lines low over sea, showing no wing markings.

JUVENILE Resembles female, cheeks less marked.

HABITAT Breeds beside lakes and rivers on moorland and tundra in N and NW Europe, winters at sea on Atlantic coasts.

NEST Well-concealed, down-lined cup on ground.

VOICE Croons and growls on breeding grounds.

GENERAL Regular, locally common. Scarce Velvet Scoter (*M. fusca*): similar in most respects, but in flight both sexes show bold white patch on wing.

Goldeneye

Bucephala clangula

ADULT Medium (48 cm) sea duck, groups often dive in unison. Drake has dark head, white face spot. Duck brown above, dark brown head and white belly. Watch for bulky, angular head profile in both sexes. In flight, whirring noisy wingbeats and white wing patches are conspicuous.

JUVENILE Resembles female, as does eclipse male.

HABITAT Summer visitor or year-round resident in N Europe, winter visitor elsewhere. Favours marshy forests for breeding; winters at sea or on larger fresh waters.

NEST Down-lined cup in old burrow or hollow tree; uses nestboxes well above ground level.

VOICE Rarely vocal; nasal quacks or low growls.

adult female

GENERAL Regular in both winter and summer, but never numerous.

adult male

Clangula hyemalis

ADULT Distinctive medium (50 cm) sea duck – but one third of this is tail. Winter drake is largely white with brown patches (note two-tone beak); summer drake largely chocolate brown with white cheeks and flanks. Duck in summer is brown above, white below, with white face patches; winter duck has more white on head and neck.

JUVENILE Resembles winter female.

HABITAT Breeds beside lakes in tundra of far N Europe, wintering on N and W coastal seas, very occasionally on larger fresh waters inland.

NEST Well-concealed, down-lined grass cup on ground.

VOICE Noisy; high-pitched goose-like honks.

GENERAL Locally regular, in places fairly common.

adult male winter

adult male summer

adult female winter

ADULT Large (60 cm), sea duck with wedge-shaped head profile. Dives for shellfish. Drake strikingly pied in flight, with black belly; duck well camouflaged in browns, shows white underside to forewing. Immature drakes blotched black and white. Flies heavily and low over the sea.

JUVENILE Resembles female; eclipse and young males gradually acquire white plumage.

adult male

adult female

HABITAT Breeds on N coasts, winters S to Biscay.

NEST Grassy cup with copious downy lining, on ground.

VOICE Vocal; drake uses loud moaning crooning, duck harsh *corrr*.

GENERAL A typical N coastal bird; fairly widespread and locally common.

DISTRIBUTION
Scarce resident;
wild birds
sometimes visit
from mainland
Europe

adult
male

ADULT Large (55 cm) diving duck, but dives
infrequently and behaves like a surface-feeder. Drake
striking, duck brown, paler on belly; watch for dark
brown crown contrasting with distinctive pale grey
cheeks. In flight, both sexes show bold broad white
wingbar running the length of the wing.

JUVENILE Resembles female, as does eclipse male,
but pale cheeks less prominent.

HABITAT Year-round resident and winter visitor to
reed-fringed fresh or brackish wetlands in S Europe,
scarce visitor or vagrant further N.

NEST Well-concealed, down-lined grass cup on ground.

VOICE Harsh *kurr*.

GENERAL Uncommon, sometimes plentiful in winter.

adult
female

adult male

ADULT Medium (45 cm) diving duck. Watch for characteristic wedge-shaped head profile. Drake sombre but distinctive, in grey, black and chestnut; duck rufous-brown above, paler on face, throat and belly. In flight, both sexes show greyish wings with indistinct paler grey wingbars.

JUVENILE Resembles female, as does eclipse male.

HABITAT Summer visitor to N and E Europe, resident year-round in some W areas, winter visitor further S.

NEST Well-concealed, down-lined grass cup on ground.

VOICE Rarely vocal; duck uses hoarse growl in flight.

adult female

GENERAL Widespread, locally common.

Mergus albellus

ADULT Small sawbill duck (41cm). Adult male white with a large head and small, silver beak; narrow black lines over back and a black mask over the eyes. Black markings are much more striking in flight. Female grey with maroon head and white cheeks.

JUVENILE As adult female.

HABITATS Breeds by lakes and ponds in northern forests.

NEST Tree hole lined with feathers and down.

VOICE Grunts while courting.

adult
male

adult female

GENERAL Occasional visitor to Britain in winter. Female may be confused with Red-crested Pochard *(Netta rufina)* or Slavonian Grebe *(Podiceps auritus)*.

ADULT Large (55 cm) sawbill duck with slim, cigar-shaped body, long slim beak (with serrated edges to grip fish) and untidy bristling crest. Dives frequently. Drake subtly elegant, duck has brown head. In flight, both sexes look elongated, showing white patches on inner wings.

JUVENILE Resembles dull female, as does eclipse male.

HABITAT Summer visitor to far N Europe, year-round resident in W. Breeds along coasts and beside fast-moving fresh waters; winters in similar areas, at sea, and on larger inland fresh waters such as reservoirs.

NEST Down-lined in burrow or hollow.

VOICE Normally silent.

GENERAL Fairly widespread, regular, but rarely numerous.

adult male

adult female

adult female

adult male

ADULT Large (65 cm) sawbill duck, slim-beaked with a streamlined, cigar-shaped body. Dives frequently. Drake strikingly white at a distance, often tinged pink at close range. Both sexes have bulky but smooth crests, giving angular head profiles. Duck has silver-grey body, white on belly, and chestnut head. Both show white on inner wing in flight.

JUVENILE Resembles dull female, as does eclipse male.

HABITAT Summer visitor to far N Europe, wintering in NW. Breeds beside fast-moving rivers, winters on larger fresh waters inland, rarely on salt waters.

NEST Down-lined, in burrow or hollow tree. Normally silent.

GENERAL Fairly widespread, regular, but rarely numerous.

ADULT Large (52 cm), buzzard-like raptor. Plumage variable. Watch for small grey head with inconspicuous beak and, in flight, boldly barred underwing and black 'wrist' patches. Tail usually held closed, looking long and narrow with three bold bars, one at tip, two near base. Sexes similar.

JUVENILE Variable (as adult), but browner.

HABITAT Summer visitor to forests and woodland over much of Europe except extreme W; migrant or vagrant elsewhere.

NEST Bulky structure of sticks high in tree.

VOICE Rapid squeaky *kee-kee* or *kee-aa*.

GENERAL Widespread, but never numerous. Usually solitary, but occurs in groups along migration routes.

adult

Hen Harrier

Circus cyaneus

adult male

adult female

ADULT Medium (48 cm), long-winged raptor. Watch for large, white rump patch on generally brown female. Male pale grey with white rump and black wing tips. Hunts low, gliding on stiff wings held in shallow V, tail long, usually held unfanned, looking narrow.

JUVENILE Browner than female, more dark streaks.

HABITAT Summer visitor to N and NE Europe, year-round resident or winter visitor elsewhere. Breeds in dense ground vegetation (eg heather). Favours open landscapes including moorland, young forestry plantations, marshland (especially in winter). Roosts communally in winter.

NEST Rough grassy platform on ground.

VOICE Chattering *kee-kee-kee*, but rarely vocal.

GENERAL Widespread, rarely numerous.

ADULT Large (53 cm), broad-winged raptor; male brown above, chestnut below, with distinctive long grey tail and grey, brown and black wing pattern in flight. Female has uniformly rich brown body and wings, with pale creamy-white crown and throat. Usually hunts low, gliding over reedbeds, wings held stiffly in a shallow V.

JUVENILE Paler brown with heavy darker streaks.

HABITAT Summer visitor to central and E Europe, year-round resident in S. Favours extensive wetlands with reedbeds.

NEST Platform of reeds deep in reedbed.

VOICE Rarely heard **kee-yah**.

GENERAL Broader-winged and more small-eagle-like than most harriers. Widespread, locally quite common.

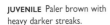

adult male

adult female

ADULT Large (62 cm), long-winged raptor with long, deeply forked, reddish tail distinctive in flight. Watch for pale head when perched. In flight, underwings show large whitish patches near tips, contrasting with black primary feathers. Often soars. Sexes similar.

adult male

JUVENILE Resembles adult, but duller and browner.

HABITAT Widespread summer visitor to central Europe, year-round resident in S regions and extreme W (Wales). Favours open woodland and farmland, often in hills.

NEST Bulky, untidy structure of twigs etc., high in a tree.

VOICE Repetitive buzzard-like **tee-tee-teear**.

GENERAL Though widespread, usually solitary and rarely numerous.

Montagu's Harrier

Circus pygargus 63

ADULT Medium (40 cm), but small and slim for a harrier. Male like Hen Harrier (p.60), but note black bar on trailing edge of wing, chestnut streaks on flanks and no white rump. Female similar to female Hen Harrier, but white rump patch smaller; has distinctive owl-like face markings. Flight more buoyant than other harriers, wings narrower and more pointed.

JUVENILE As female, but richer-brown, less boldly marked.

HABITAT Summer visitor to S and central Europe; breeds in open habitats from farmland to marshes, moors and sand-dunes.

NEST Grassy platform on ground.

VOICE Shrill *keck-keck-keck*.

GENERAL Widespread, but erratic, rarely common.

adult male

adult female

ADULT Large (55 cm), heavily-built, but fast-flying hawk, larger female almost buzzard-sized. Watch for prominent eyestripes meeting on nape to give capped appearance, and fluffy white undertail coverts. In flight, note long, broad rounded wings and long tail. Usually among trees, but soars in tight circles at height in spring.

juvenile

JUVENILE Resembles adult, but browner and more scaly above, buff below with heavy streaking.

HABITAT Year-round resident over much of Europe. Favours extensive forests or woodlands.

NEST Platform of twigs high in tree.

VOICE Chattering *kek-kek-kek* or geck.

GENERAL Widespread, but generally scarce. Easiest seen in spring when displaying.

ADULT Medium (35 cm), fast-flying hawk. Watch for short, rounded wings, long four-barred tail and distinct eyestripes. Upright perching stance. Smaller male greyish above, barred reddish on breast, larger female barred dark brown on breast, grey-brown back.

JUVENILE As female, but streaked (not barred) on breast.

HABITAT Year-round resident over most of Europe, summer visitor to far N. Favours farmland with trees and woodland/forest of all types.

NEST Platform of twigs, high in tree.

VOICE Sharp, fast *keck-keck-keck*.

GENERAL Widespread, locally fairly common, increasing after recent pesticide-induced drastic decline.

adult

juvenile

Buteo buteo

ADULT Large (55 cm) raptor with long, broad, heavily fingered wings. Plumage very variable, usually dark brown above, paler with dark streaks below. Soars: watch for short fanned and rounded tail, black patches in paler areas of underwing at 'wrist'. Sexes similar.

JUVENILE Similar to adult.

HABITAT Year-round resident over much of Europe, summer visitor to far N. Favours open country including mountains and moorland, often with tracts of woodland.

NEST Bulky twig structure in tree or occasionally on ground.

VOICE Distinctive far-carrying cat-like mewing.

GENERAL Widespread, locally fairly common. Perches solid and upright, on posts or telegraph poles. Visits carrion.

adult
pale

adult
dark

adult

ADULT Huge
(85 cm) raptor with big beak
and long, broad, heavily fingered
wings. Adult brown, golden feathers on head and neck visible at
close range. Soars frequently; watch for long, broad tail,
prominent head, and wings held slightly above horizontal with
upcurled tips. Sexes similar.

JUVENILE As adult, but with white patches in wing and black-
tipped white tail.

HABITAT Remote and extensive mountain areas, often with
forest, throughout Europe, down
to sea level in N.

NEST Enormous structure of
branches, used year after year, in
tree or on high rocky ledge.

VOICE Rarely vocal; *kaah*.

GENERAL Though widespread,
always scarce. Often confused
with much smaller Buzzard (p.66).

Peregrine

Falco peregrinus

ADULT Medium (45 cm), but largest European falcon. Grey above, with finely, dark-barred, white underparts; black moustache and white cheeks. In flight, watch for relatively short, broad-based pointed wings. Circles high waiting for prey to fly below, then plunges at high speed in pursuit (stoop).

adult

JUVENILE Brown and scaly above, buff below with dark streaks not bars.

HABITAT Year-round resident or winter visitor over much of Europe, summer visitor to far N. Breeds on mountains, moors or coasts with cliffs; winters on moors and coastal marshes.

NEST Eggs laid in bare scrape on cliff ledge.

VOICE Harsh *keck-keck*.

GENERAL Widespread, never numerous, but range and numbers increasing.

ADULT Large (58 cm) brown and white raptor. Watch for pale underparts and white head with dark eye patches. Carries wings in a distinctive open M in flight, hovers clumsily over water, then plunges spectacularly for fish prey. Sexes similar.

JUVENILE As adult, but less distinctly marked.

HABITAT Summer visitor breeding beside N European lakes, rivers and coasts; also year-round resident or winter visitor to extreme S Europe. Migrates over almost any water.

NEST Bulky, of branches, usually in tree.

VOICE Rarely vocal; whistling *tchew*.

adult

GENERAL Widespread, but never numerous.

Falco tinnunculus

adult male

adult female

ADULT Medium (35 cm) falcon, long-winged and long-tailed, distinctively hovers before diving onto prey. Female brown with multi-barred tail; male has black-spotted chestnut back, grey head, grey tail with black terminal bar. Usually solitary.

JUVENILE As female, but duller.

HABITAT Widespread and common resident year-round except in far N, where is summer visitor. Almost any habitat.

NEST Lays eggs on bare ledge or in hole.

VOICE Shrill *kee-kee-kee*, usually when breeding.

GENERAL Lesser Kestrel *(F. naumanni):* gregarious, locally common summer visitor to S Europe, breeding colonially. Male has unspotted chestnut back and pale appearance; rarely hovers.

ADULT Medium (30 cm) falcon, compact and low-flying, catches prey (usually birds) by surprise and speed. Watch for dark slate-grey back and tail of male, female dark brown, paler below, copiously streaked. In flight, wings heavily barred on undersides, powerful and pointed; tail long, multi-barred in female, with single terminal bar in male.

JUVENILE As female, but more rufous.

HABITAT Summer visitor to N Europe, winter visitor or year-round resident elsewhere. Breeds on moors, tundra and rough grassland, often winters on coastal marshes.

NEST Shallow depression on ground.

VOICE Chattering *kee-kee-kee*.

GENERAL Widespread, but always scarce. Can occur almost anywhere on migration.

adult male

adult female

Hobby

Falco subbuteo

ADULT Medium (28 cm), fast-flying falcon. Watch for distinctive flight silhouette like giant swift with long sickle-shaped wings. White collar prominent at a distance, black moustaches, heavily barred underwings and chestnut undertail only clear at close range.

JUVENILE Brown, not grey above; buff, with dark streaks below, lacking chestnut.

HABITAT Summer visitor to European heathland and farmland except in far N. Often hunts over water.

NEST Usually lays in abandoned crow's nest.

VOICE Sharp *kew* and repetitive *ki-ki-ki*.

GENERAL Widespread, locally fairly common especially in warmer S areas with plentiful large insect prey.

adult

Distribution Circumpolar, through most of Sacndinavia except lowland areas

ADULT Medium (40 cm), well-camouflaged, heavily-built game bird. Summer male mottled dark chestnut, red wattles over eyes, striking white wings in flight. Female also white-winged, duller and greyer, no wattles. Both sexes are white in winter except black tail. Remains still until danger close, then whirrs off on noisy, downcurved wings.

JUVENILE Much as female.

HABITAT Year-round resident of moorland, tundra and birch scrub across N Europe.

NEST Well-concealed grassy cup on ground.

VOICE Loud and distinctive *go-back-urrr*.

GENERAL Locally common, as is British and Irish subspecies, Red Grouse: lacks white plumage year-round.

adult

adult male winter

ADULT Medium (35 cm), high-altitude game bird. Summer adult richly mottled brown and grey above, with white belly and wings. In winter, largely white except black tail. Watch for distinctive dark mark through eye and feathered feet, well insulated from snow. Usually run from danger rather than flying. Sexes similar except male has red wattles over eyes.

JUVENILE As summer adult, but duller.

HABITAT Year-round resident of N tundra and isolated mountain areas elsewhere in Europe.

NEST Well-concealed grassy cup on ground.

VOICE Croaking *arr-arr-kar-kar-kar*.

GENERAL Restricted by habitat, rarely numerous.

ADULT Large (50 cm), bulky game bird. Male glossy black with white wingbar and white underside to lyre-shaped tail. Bright red wattles over eyes. Female mottled grey-brown, with longish slightly forked tail. Gathers at dawn and dusk on communal display grounds (leks). Flies high, fast and far when disturbed.

adult female

JUVENILE As female.

HABITAT Year-round resident across N and NW Europe. Favours heaths, rough grass, moorland and open woodland. Locally common.

NEST Well-concealed grass cup on ground.

VOICE Cacophonic croons and bubblings at lek.

GENERAL Capercaillie (*T. urogallus*): black turkey-like male (85 cm) with shaggy throat feathers and white beak. Female smaller and chestnut.

adult male

ADULT Small (18 cm), vocal, but secretive and well-camouflaged game bird, smaller than a thrush. Underparts sandy buff, upperparts mottled browns, fawns and chestnuts, white streaked. Watch for broader buff stripes on crown and small black bib of male, otherwise sexes similar. Flies only as last resort.

JUVENILE As adult, but duller.

HABITAT Summer visitor to much of Europe except far N, sometimes year-round resident in extreme SW. Favours open farmland, grassland and heath.

NEST Well-concealed grassy cup on ground.

VOICE Ventriloquial *wet-my-lips* call.

GENERAL Widespread, variable from year to year, rarely numerous.

Red-legged Partridge

ADULT Medium (35 cm), dumpy, upright game bird. Black-bordered white bib and speckled gorget; white eyestripe and striking black, white and chestnut bars on flanks. Plain rich brown back. Beak and legs deep pinkish red. Sexes similar.

JUVENILE Sandy brown, lacks head and flank patterns.

HABITAT Year-round resident in W and SW Europe. Favours drier farmland, heath, downland and scrub.

NEST Well-concealed grassy cup on ground.

VOICE Distinctive *chuck, chuck-arr*.

GENERAL Locally common, often in small flocks. Sometimes artificially introduced. Rock Partridge (*A. graeca*) of rocky Mediterranean hillsides is similar, as is Chukar (*A. chukar*) of extreme SE. Both lack the speckled gorget.

adult

adult

ADULT Medium (30 cm), dumpy, upright game bird. Finely mottled and white-streaked grey, buff and chestnut upperparts, grey breast and dark brown horseshoe mark on belly. Erratically barred chestnut flanks. Crouches, well-camouflaged, flying at last moment. Whirrs off fast and low on downcurved wings, shows chestnut sides to tail. Sexes broadly similar: female has less marked horseshoe.

JUVENILE Sandy and streaked brown, lacks marks.

HABITAT Year-round resident over much of Europe except N and SW. Favours farmland, grassland, heath and scrub.

NEST Well-concealed grassy cup on ground.

VOICE Distinctive, rusty *kirrrr-ick*.

GENERAL Widespread, once fairly common, but locally declining. Often in small flocks (coveys).

adult

adult

ADULT Large (85 cm, but half is long tail) distinctive game bird. Male unmistakable, rich gold and chestnut, with green head and scarlet, fleshy face patch. Female smaller, well camouflaged in mottled browns and buffs, with long central tail feathers.

adult female

adult male

JUVENILE As female, but shorter-tailed.

HABITAT Introduced centuries ago, widespread year-round resident across much of Europe except far N. Favours farmland, heath, scrub and open woodland.

NEST Well-concealed grassy cup on ground.

VOICE Ringing, far-carrying **kok-kok**, followed by explosive wing claps.

GENERAL Numbers variable as stocks artificially augmented for shooting; locally common.

Rallus aquaticus

adult

ADULT Medium (28 cm) skulking crake. Watch for dull grey breast, black-barred flanks, and white underside to frequently flicked tail. Beak long, downcurved, deep red with black tip. Long legs and spidery toes pinkish. Slips silently through reeds. Flies weakly and low, legs trailing. Sexes similar.

JUVENILE Darker and duller, more speckled and barred than adult.

HABITAT Widespread across much of Europe year-round, summer visitor in N. Favours densely vegetated wetlands, swamps and reedbeds.

NEST Well-concealed cup on ground, deep in cover.

VOICE Often noisy; pig-like grunts and squeals.

GENERAL Difficult to see; commoner than it seems.

ADULT Stocky water bird (27cm), spends most of its time hidden. Resembles other rails but wings are rich chestnut and head and front of neck are grey. Bill is short and pale. Chestnut wings particularly noticeable in flight. Females show less grey.

JUVENILE Shows no grey; reddish flanks

HABITAT Low, wet marshland, and high grass and hay meadows.

NEST Flat, made of grass and hidden in vegetation

adult

VOICE Loud, grating '***crek-crek***', heard early morning and late evening

GENERAL Now very rare and very hard to see, does breed Ireland and west of Scotland. Mostly known by its call.

DISTRIBUTION
Breeds across N
Palearctic, and
migrates through
Holland and
France to Spain

ADULT Huge
(110 cm), stork-
like, long-legged,
long-necked
marsh bird.
Watch for black and white neck
markings, inconspicuous red
crown, and bulky, bushy plumes
over tail. Flies neck and legs extended,
usually in V-formation, often calling.
Sexes similar.

adult

JUVENILE Grey brown, paler on underside, lacking head markings
and plumes.

HABITAT Summer visitor breeding on far N marshes and tundra;
migrates via established staging posts on farmland or marshland.

NEST Huge reed platform rising above swamp.

VOICE Fabulous wild trumpeting in flight, whooping calls on
breeding grounds.

GENERAL Generally scarce, flocks on migration.

adult

ADULT Medium (33 cm) familiar crake. Watch for dull-black plumage with white flank streak and white underside to frequently flicked tail; short yellow beak and red fleshy forehead shield. Legs greenish with distinctive red 'garter', toes spidery. Sexes similar.

JUVENILE Nestling has black fluffy down; immature brown above, fawn below, lacks frontal shield.

HABITAT Year-round resident over most of Europe, summer visitor to N and NE. Favours fresh waters from smallest pond to largest lake.

NEST Bulky mound of waterweed, often in or over water.

VOICE Varied ringing calls, including *whittuck*.

GENERAL Widespread, often common.

Fulica atra

adult

ADULT Medium (38 cm), familiar, dumpy all-black rail. Watch for grey-green legs with lobed toes, white beak and frontal shield. Aggressive, often fluffs out feathers and fights. Flies low, legs trailing, showing white trailing edge to wing. Dives frequently. Sexes similar.

JUVENILE Downy young fluffy and black; immature dark grey above, whitish throat and belly, lacks frontal shield.

HABITAT Year-round resident over much of Europe, summer visitor to N and NE. Favours larger fresh waters, occasionally on sheltered estuaries.

NEST Conspicuous mound of waterweed, usually in or over water.

VOICE Metallic and strident *kook* or *kowk*.

GENERAL Widespread, conspicuous and generally common.

ADULT Large, brown wader (41cm). Large head and yellow, staring eye. Long yellow legs, and yellow bill with black tip. Strong, slow flight, strong black and white wing pattern. Well-camouflaged against stony ground, will stand very still.

JUVENILE Less obvious wing bar, markings generally less well-defined.

HABITAT Open ground with sparse vegetation.

NEST Eggs laid on bare scrape.

VOICE Whining *coor-lee*.

GENERAL Summer visitor in Britain, winters in SW Europe and Africa.

adult

ADULT Medium (43 cm), but among the larger, more robust waders. Watch for strikingly pied plumage, stout, straight orange beak and thick, fleshy pink legs. In winter, has inconspicuous white collar. In flight shows bold white wingbars and black and white rump and tail pattern. Sexes similar.

JUVENILE Dull, sooty version of adult.

HABITAT Year-round resident, usually coastal, but also on damp meadows in W Europe; summer visitor breeding on N coasts and marshes; winter visitor to S shores.

NEST Shallow scrape lined with pebbles, seaweed or grass.

VOICE Strident pipings and *kleep* calls.

adult

GENERAL Comparatively widespread, often common; usually in flocks, sometimes large.

Black-winged Stilt

Himantopus himantopus

adult

ADULT Medium (38 cm) wader, unmistakable if full length of pink legs can be seen. Watch for jet black back and wings contrasting with white body and needle-slim, straight black beak. In winter has smoky crown and nape. Flies with long legs trailing distinctively. Sexes similar.

JUVENILE Duller and browner than adult, with long brownish legs.

HABITAT Summer visitor (occasionally year-round resident) in extreme S Europe; vagrant elsewhere. Breeds by saltpans, brackish lagoons and freshwater marsh pools.

NEST Shallow scrape on ground, lined with pebbles, shells or fragments of vegetation.

VOICE Vocal; strident yelping *kyip*.

GENERAL Locally fairly common, often in flocks.

Recurvirostra avosetta

adult

ADULT Medium (43 cm), unmistakable wader. Watch for boldly patterned pied plumage and unique long, finely-pointed, upturned black beak. Legs long, blue grey. Feeds by sweeping beak from side to side through shallow water. Sexes similar.

JUVENILE Greyer version of adult.

HABITAT Summer visitor to a few N coastal marshes, winter visitor to others and to sheltered estuaries, year-round resident on S marshes, saltpans and lagoons.

NEST Shallow scrape on ground, lined with fragments of shell or nearby vegetation.

VOICE Vocal; distinctive **kloo-oot** and **kloo-eet**.

GENERAL Locally common, generally increasing. Breeds colonially, often feeds in flocks in winter.

ADULT Small (15 cm), lightweight, fast-moving plover. Watch for short, slim black beak, slender black legs, black and white forehead, pale chestnut crown. Black marks on shoulders. Female and winter birds paler, lack chestnut cap. In flight shows white sides to tail and white wingbar.

JUVENILE Scaly, sandy back; inconspicuous head and collar markings.

HABITAT Year-round resident or winter visitor to Mediterranean coasts, summer visitor to W European coasts. Favours saltpans, muddy lagoons and sandy beaches.

NEST Shallow scrape lined with shell fragments.

adult

VOICE Melodious **choo-wit**, soft **wit-wit-wit**.

GENERAL Locally common in S, scarcer on W coast, vagrant elsewhere.

Charadrius hiaticula

adult

ADULT Small (20 cm), fast-moving plover. Watch for stubby black-tipped orange-yellow beak, black and white forehead, broad, black collar band, and orange legs. In flight shows broadly white-bordered tail and striking white wingbars. Sexes similar.

JUVENILE Sandier and scaly above, lacks black markings, has brown collar.

HABITAT Summer visitor to N coasts, year-round resident on W coasts; winter visitor to S shores. Favours sandy coasts and saltpans, occasionally inland excavations.

NEST Shallow scrape lined with fragments of local material, on ground and well camouflaged.

VOICE Melodious **too-lee**; trilling song near nest.

GENERAL Widespread and fairly common.

Little Ringed Plover

Charadrius dubius

adult

ADULT Small (15 cm), fast-moving plover. Short black beak and complex head pattern, with white stripe between crown and black face and forehead bar; yellowish legs and yellow eye-ring. In flight shows slim black collar, white nape, white sides to tail and lack of wingbar. Sexes similar.

JUVENILE Scaly brown above, with greyish collar, pale yellow eye-ring.

HABITAT Summer visitor to most of Europe. Favours sandy coasts, lagoons and saltpans, and inland sandpits, quarries and other excavations.

NEST Shallow scrape on ground, lined with fragments of local material. Very well concealed.

VOICE Quiet, piping **tee-you**; trills near nest.

GENERAL Widespread, locally fairly common in S, scarce elsewhere.

Pluvialis apricaria

ADULT Medium (28 cm), but largish among plovers. In summer, watch for striking black-speckled gold back and white-bordered black belly. In winter, dull brown-speckled golden buff upperparts, buff breast and white belly. Flight swift, showing indistinct wingbar. Sexes similar.

adult non-breeding

JUVENILE As winter adult, but drabber.

HABITAT Summer visitor breeding on N moorland and tundra, winters on W wet grassland, farmland and marshes.

NEST Well-concealed grass-lined scrape on ground.

VOICE Sad but melodious *tloo-eee*.

GENERAL Widespread, but not numerous as a breeding bird, regular and locally fairly common in winter, often in large flocks.

ADULT Medium (28 cm), bulky plover. In summer, black-speckled silver back and white-bordered striking black belly. In winter, dull blackish-speckled grey back and white underparts. In flight watch for black 'armpits', white wingbar, white rump and dark-barred tail. Sexes similar.

JUVENILE As winter adult, but drabber.

HABITAT Scarce summer visitor breeding in extreme N; more numerous on migration or as winter visitor to estuaries and sheltered sandy or muddy coasts along Atlantic and Mediterranean seaboards.

adults
non-breeding

NEST Shallow scrape on tundra.

VOICE Plaintive *tee-loo-eee*.

GENERAL Widespread, but rarely numerous, often solitary.

Vanellus vanellus

ADULT Medium (30 cm), distinctive plover, predominantly black and white, at close range black areas shot with iridescent purple and green. Watch for conspicuous long slender black crest, chestnut undertail, floppy flight on markedly rounded black and white wings. Sexes broadly similar.

JUVENILE Drabber, scaly-backed with short crest.

HABITAT Summer visitor to N and E Europe, year-round resident, migrant or winter visitor elsewhere. Breeds on fields, moorland and marshes; winters on similar areas, also estuaries and arable farmland.

NEST Grass-lined scrape on ground.

VOICE Very distinctive *pee-wit*.

adult

GENERAL Widespread, often common, often in substantial flocks.

ADULT
Small–medium
(25 cm) wader.
Summer adult
chestnut with mottled gold and brown back. In winter, lacks
distinction: dull grey, with medium beak, medium legs and
indistinct wingbar. Watch for bulky build and whitish eyestripe.
Sexes similar.

JUVENILE As winter adult, but browner on back.

HABITAT Breeds on Arctic tundra. Brief
spring migration up European coasts,
but most in autumn or winter. Favours
estuaries and sheltered sandy or muddy bays.

adult
breeding

NEST Well-concealed grass-lined scrape on
ground.

VOICE Occasional but distinctive grunt.

GENERAL Regularly in enormous flocks. Packs close together
on ground and in flight, wheeling and turning as one.

ADULT Small (20 cm), fast-running wader. Summer adult rich rufous cinnamon. In winter, distinctively pale: watch for short dark beak, black smudge through eye. In flight shows bold white wingbar. Chases waves in and out on the sand.

JUVENILE As winter adult, but with brown mottled back.

HABITAT Breeds on Arctic tundra. Brief spring migration on W coasts, but most in autumn or winter on sandy beaches and bays.

NEST Well-concealed grassy cup on ground.

VOICE Repeated short sharp *quick*.

GENERAL Widespread and regular, locally fairly common though rarely in large flocks.

adult breeding

**adult
breeding**

ADULT Tiny (13 cm) wader. Short fine beak, black legs and grey outer tail feathers in flight. In summer, rich mottled brown above, with double V marking on back. In winter, pale dull grey, with traces of the V. Sexes similar.

JUVENILE Much as summer adult.

HABITAT Migrant or summer visitor breeding on far N tundra. Some overwinter in W and S. Favours saline lagoons and creeks, freshwater pools and swamps.

NEST Well-concealed grass-lined cup on ground.

VOICE Terse *chiff*.

GENERAL Widespread, but rarely numerous. Similar-sized but drabber Temminck's Stint *(C. temminckii)* scarcer, yellow legs, white outer tail feathers and rattling *tirrrr* call.

Curlew Sandpiper

WADERS

Calidris ferruginea

adult
non-
breeding

ADULT Small (20 cm) wader with distinctive downcurved beak. Summer adult mottled brown above, rich chestnut below. Winter birds pale grey above, white below. Watch for eyestripe and, in flight, square white rump contrasting with black tail. Sexes similar.

JUVENILE As winter adult, slightly buffer, with scaly pattern on back.

HABITAT Summer visitor breeding on Arctic tundra. Regular spring and autumn migrant and occasional overwintering visitor to W and S coasts. Favours sheltered bays, estuaries, lagoons and freshwater marshes.

NEST Well-concealed grassy cup on ground.

VOICE Distinctive trilling **chirrup**.

GENERAL Regular, but yearly numbers vary greatly.

ADULT Small (21 cm), squat wader. In summer, mottled chestnut above, pale with dark markings below. In winter, dull, purplish, leaden-grey above, paler and grey-spotted below. Short yellow legs, black-tipped yellow beak, and white edges to black rump and tail in flight. Sexes similar.

JUVENILE As winter adult, but with some darker mottling on back.

HABITAT Breeds on Arctic tundra, winters (immature birds may be present all year) on rocky W and NW coasts.

NEST Well-concealed grassy cup on ground.

VOICE Normally silent, occasional *wit-wit*.

GENERAL Scampers among breaking waves on seaweed-clad rocky shores. Inconspicuous but approachable. Regular and locally fairly common.

adult
breeding

ADULT Small (18 cm) wader with long downcurved beak. Summer adult speckled bronze and chestnut above, white below with distinctive black belly patch. In winter, nondescript grey above, white below. In flight, watch for white wingbar and white-edged black rump and tail. Sexes similar.

JUVENILE As winter adult, but generally buffer, with brown mottling on back.

HABITAT Summer visitor or year-round resident breeding on N tundra, moorland and coastal marshes. Winter visitor or migrant to lagoons, estuaries and sheltered bays along entire European coast, inland on marshes.

adult
non-breeding

NEST Well-concealed grassy cup on ground.

VOICE Purring trill in flight; nasal *shreeep* call.

GENERAL Common, often numerous, often in flocks.

ADULT Medium (35 cm), bulky, long-beaked wader, with mottled plumage. Pale forehead and cross-wise buff stripes on angular head. Note large eyes, rounded wings and moth-like flight showing no wing markings. Sexes similar.

JUVENILE As adult, but duller.

HABITAT Summer visitor to N Europe, year-round resident or winter visitor to S. Aberrant for a wader in favouring damp woodland year-round.

adult

NEST Scrape in leaf litter on ground.

VOICE Usually silent, but in evening calls a frog-like **orrrt-orrrt** and a sneezing high-pitched **tswick**.

GENERAL Widespread and regular, difficult to see as relies on camouflage until danger very close.

ADULT Small, brown wader (19cm). Well camouflaged in brown, black and creamy-white; black on crown and in stripes on head. Bill only slightly larger than the head. Underparts white with dark streaking. Brown and white streaking on the breast. Flies only rarely when disturbed; underwings half black and white.

JUVENILE As adult

HABITAT Breeds in bogs in northern Europe; winters around muddy pools and inland marshes

NEST Cup-shaped, in grass

VOICE Usually silent

GENERAL Smaller than Common Snipe (*Gallinago gallinago*), with shorter bill

adult

adult

ADULT Medium (28 cm), squat, well-camouflaged wader with long beak. Longitudinal buff stripes on crown and back. Legs green, relatively short. Zig-zag flight. Sexes similar.

JUVENILE As adult, but duller.

HABITAT Summer visitor to N Europe, year-round resident in central and W areas, winter visitor further S. Favours saline lagoons, swamps, reedbeds and wet grassland.

NEST Well-concealed deep cup in grass.

VOICE Repeated **tick-er** in breeding season, harsh scarp when flushed. Tail feathers produce vibrant 'drumming' noise in diving display flight.

GENERAL Widespread, local. Smaller, scarcer, shorter-beaked Jack Snipe *(Lymnocryptes minimus)*: difficult to flush, silent, flies low and straight for short distance, showing no white on tail.

Philomachus pugnax

ADULT Medium (30 cm), long-legged wader. Larger summer male has large bright ruff of feathers. Female and winter male duller and scaly. In flight, watch for white oval patches on each side of rump and tail; long, usually orange, legs extend beyond tail. Beak orange and black in male, blackish in female.

JUVENILE Buff head and neck, scaly brown back, white belly.

HABITAT Summer visitor to N Europe, breeding on tundra and marshland; year-round resident or winter visitor in S. Favours coastal lagoons or inland marshes.

NEST Well-concealed deep cup in grass.

VOICE Usually silent, sometimes *chuk-uk*.

GENERAL Widespread, but scattered, rarely numerous.

adult male

adult female

ADULT Medium (40 cm), but large and tall for a wader. Note long, straight beak. In summer, has chestnut head and neck; in winter, dull pale grey with white belly. In flight, watch for striking black and white wingbars, white rump and black tail. Sexes similar.

JUVENILE As winter adult, but buffer, with scaly back pattern.

HABITAT Summer visitor or year-round resident to north-central and NW Europe, winter visitor to S coasts. Breeds on damp grassland and marshes, winters on sheltered estuaries and bays.

adult

NEST Well-concealed cup deep in tussock.

VOICE Noisy *wicka-wicka-wicka* when breeding.

GENERAL Locally fairly common, may flock in winter.

Limosa lapponica

ADULT Medium (40 cm), but large and long-legged for a wader. Note long, slightly upturned beak. In summer, bright chestnut with brown-mottled back; in winter, brown with darker streaks above, whitish below. In flight, note lack of wingbar, white rump and dark-barred white tail. Sexes similar.

JUVENILE As winter adult, with chestnut mottling on back.

HABITAT Summer visitor breeding on Arctic tundra, migrant or winter visitor to W and S European coasts. Favours estuaries and sheltered muddy or sandy bays.

NEST Well-concealed scrape on ground.

VOICE Rare; harsh *kirrick* when breeding.

GENERAL Widespread and gregarious, locally common, often in large flocks.

adult

adult

ADULT Medium (40 cm), mottled brown, long-legged, long-necked wader. Long, downcurved, black beak appreciably shorter than Curlew (p.108). Watch for broad buff stripes on brown crown. In flight, note lack of wingbar, white rump and barred tail. Sexes similar.

JUVENILE Resembles adult.

HABITAT Summer visitor breeding on N moors, marshes and tundra, migrant (often, but not always, on coast) elsewhere.

NEST Well-concealed grassy cup on ground.

VOICE Far-carrying piping whistle, several times in rapid succession: *pee-pee-pee-pee-pee-pee-pee*.

GENERAL Widespread, regular on migration, rarely numerous.

Numenius arquata

adult

ADULT Large (58 cm), mottled brown and buff, long-legged, long- necked wader. Extremely long, downcurved beak. In flight, note lack of wing markings, pale rump extending well up back, and narrow, dark-barred white tail. Sexes similar.

JUVENILE Resembles adult.

HABITAT Summer visitor to N and north-central Europe, year-round resident, migrant or winter visitor to W and on S coasts. Breeds on moors, wet grassland and marshes; winters on sheltered sandy or muddy coasts and estuaries.

NEST Well-concealed grassy cup on ground.

VOICE *Coor-lee* at all times; bubbling song in display flight over breeding territory.

GENERAL Widespread; solitary or in flocks; locally fairly common.

adult

ADULT Medium (30 cm), long-legged, long-necked wader. In summer, unmistakable, uniformly sooty black with white spots. In winter, scaly silver-grey above, white below. In flight, watch for trailing dark red legs, lack of wingbar, white rump and barred tail. Sexes similar.

JUVENILE As winter adult, but buffer on shoulders.

HABITAT Summer visitor breeding on Arctic tundra; migrant or winter visitor on coasts of W and S Europe. Favours tundra areas close to tree limit, winters on marshes, estuaries and sheltered coasts.

NEST Well-concealed grassy scrape on ground.

VOICE Explosive *chew-it*.

GENERAL Regular, but rarely numerous. Often solitary.

Redshank

Tringa totanus

ADULT Medium (28 cm), wary wader. Rich brown above, streaked blackish in summer, whitish scaly marks in winter. White below, heavily brown speckled and streaked in summer, less so in winter. Note long scarlet legs, and, in flight, broad white trailing edges to wings. Sexes similar.

JUVENILE As winter adult, but buffer overall.

HABITAT Summer visitor to N and NE marshes. Year-round resident, migrant or winter visitor to W and S coasts. Breeds in marshes; winters on estuaries and coasts.

adult

NEST Concealed deep cup in grass tussock.

VOICE Stridently vociferous 'sentinel of the marshes', shrieking calls and melodious variants on **tu-lee-lee**.

GENERAL Widespread, locally fairly common; solitary, sometimes in small flocks.

adult breeding

ADULT Medium (30 cm), pale wader. Grey back with scaly pale markings in winter, with blackish blotches and streaks in summer. Head and neck pale in winter, dark-streaked in summer. Watch for thickish, slightly upturned beak and green legs; in flight shows conspicuous white rump and no wingbars. Sexes similar.

JUVENILE As winter adult, but buffer.

HABITAT Summer visitor to N tundra and moorland. Migrant or winter visitor to W and S coasts and marshes. Favours lagoons and sheltered bays and estuaries.

NEST Well-concealed scrape on ground.

VOICE Characteristically trisyllabic *tu-tu-tu*.

GENERAL Widespread, but rarely numerous.

ADULT Small (23 cm), dark wader. Dark, greenish-grey back speckled white, underparts white. Short, dark green legs. Often bobs. In flight, watch for dark underwings, lack of wingbars, bold white rump and heavily dark-barred tail. Sexes similar.

adult non-breeding

JUVENILE As adult, but more heavily speckled.

HABITAT Summer visitor to N Europe; winter visitor or migrant to coastal and inland marshes elsewhere. Breeds in swampy forest, muddy pools and creeks.

NEST Grassy scrape on ground, occasionally in deserted nest in tree.

VOICE *Tloot-weet-wit* on take-off; trilling song.

GENERAL Widespread and regular, but rarely numerous. Often solitary.

Turnstone

Arenaria interpres

adult
non-
breeding

ADULT Small (23 cm), stocky wader. Harlequin plumage, mainly browns and white in winter; black, white and chestnut in summer. Short orange legs and short, dark, wide beak (used to turn over seaweed and pebbles). In flight, watch for complex black and white pattern on back, wings and tail. Sexes similar.

JUVENILE Similar to winter adult.

HABITAT Summer visitor breeding on N rocky coasts and tundra; migrant, winter visitor (or year-round) on coasts further S. Generally maritime, favours rocky, surf-washed coasts with dense seaweed.

NEST Shallow scrape on ground.

VOICE Distinctive staccato *tuk-uk-tuk*.

GENERAL Widespread, locally common; rare inland.

Actitis hypoleucos

adult non-breeding

ADULT Small (20 cm), dumpy wader, perpetually bobbing. Sandy brown upperparts flecked with white in summer. Short greenish legs. Watch for whirring flight interrupted by glides on downcurved wings, low over water. Shows white wingbar and white edges to buff rump and tail. Sexes similar.

JUVENILE As adult, but duller and chequered.

HABITAT Summer visitor or migrant to much of Europe, winter visitor to S, may overwinter elsewhere. Breeds beside lakes, rivers and streams; winters on fresh and salt marshes, occasionally along coasts.

NEST Shallow scrape on ground close to water.

VOICE Trilling **twee-wee-wee** call; high-pitched song based on **tittyweety** phrases.

GENERAL Widespread and regular; often solitary.

Wood Sandpiper

Tringa glareola

ADULT Small (20 cm), slim-built wader. Dark grey-brown upperparts with scaly white-edged feathers. Head and neck white with darker streaks, note clear eyestripe and yellowish legs. In flight, unmarked wings with pale undersides, white rump with faintly barred tail. Sexes similar.

JUVENILE As adult, but warmer colour and more mottled.

HABITAT Summer visitor breeding on N marshes and tundra, migrant elsewhere visiting inland and coastal pools, marshes and lagoons. May overwinter in S.

NEST Shallow scrape on ground.

VOICE *Chiff-if-if* on take-off; yodelling song.

adult non-breeding

GENERAL Widespread and regular, but rarely numerous. Often solitary, occasionally in small flocks.

ADULT Small, dainty, brownish-grey wader (18cm), with a needle-sharp bill. Adult breeding birds characterised by orange-brown streaking running from behind the eye down the side of the neck. Back is streaked with the same colour. Non-breeding birds grey and white.

JUVENILE As non-breeding male, but brown and white.

HABITAT Breeds in tundra far north of Europe, rarely in north of Scotland around shallow pools, winters in Middle East

NEST Marshland tussocks.

adult non-breeding

VOICE A throaty, clucking *check* or *tyit*.

GENERAL Arrives in Britain in May on passage for summer, and returns in mid-July.

Black-headed Gull

Larus ridibundus 117

ADULT Medium (35 cm) gull. In summer, watch for dark chocolate hood, red beak and leg s. In winter, head white with black smudge behind eye. In flight shows all-black wingtips and diagnostic white leading edge to wing. Sexes similar.

JUVENILE As winter adult, but with brown W mark across wings and black tip to tail.

HABITAT Summer visitor to N and NE, year-round resident or winter visitor elsewhere. Breeds colonially on islands, dunes, sheltered coasts and beside moorland lakes. Occurs in almost any habitat.

NEST Tall mound of grass and flotsam.

VOICE Yelping *keeer* and laughing *kwaar* calls.

GENERAL Widespread, common. Scarcer Mediterranean Gull (*L. melanocephalus*): black hood and all-white wingtips, more robust.

adult winter

ADULT Medium (50 cm), gull-like seabird, slim-winged, agile in flight. Two phases: brownish-grey with darker cap or brown above with white underparts and collar. In flight, watch for long-pointed, central tail feathers, conspicuous white patches near tips of brownish wings. Sexes similar.

JUVENILE Brown, speckled and barred buff, without elongated tail feathers.

HABITAT Summer visitor breeding colonially on N coasts, islands and tundra; migrant elsewhere. Maritime, except nesting; favours inshore waters.

NEST Grass cup on ground.

VOICE Harsh *kee-aar* over colonies.

GENERAL Widespread, but numerous only near colonies. Pirates fish from other seabirds.

adult pale summer

adult dark summer

ADULT Large (60 cm) skua, uniformly dark brown flecked and streaked white and buff. In flight, is piratical (even chases Gannets (p.21)); has striking white patches at base of primaries. Sexes similar.

JUVENILE Similar to adult.

HABITAT Summer visitor breeding colonially on N islands, moors and tundra. Migrant or occasional winter visitor elsewhere. Maritime, favours coastal seas.

NEST Bulky grass cup on ground, fiercely defended.

VOICE At colony only, barking tuk or *uk-uk-uk*, nasal *skeer*.

GENERAL Locally numerous only at colonies, which are few. Regular but quite scarce except near breeding colonies.

adult

Larus fuscus

ADULT Large (53 cm) gull. Watch for yellow legs and, in flight, for black and white tips to dark grey wings. Head and neck flecked with grey in winter. Sexes similar.

JUVENILE Speckled dark brown above, paler below, all-dark primaries. Adult plumage after 3 years.

HABITAT Summer visitor breeding colonially on far N cliff-tops, islands, dunes and moors. Year-round resident or migrant along W coasts, winter visitor in S. Occurs in almost any habitat.

NEST Bulky mound of grass and flotsam on ground.

VOICE Powerful throaty *kay-ow* and laughing cries.

GENERAL Widespread, often common, increasingly wintering further N.

adult summer

Herring Gull

Larus argentatus

adult
summer

ADULT Large (55 cm) diagnostically silver-backed gull.
Watch for pink legs (but S race has yellow legs) and,
in flight, for black and white tips to pale grey wings.
Head and neck flecked with grey in winter. Sexes
similar.

JUVENILE Speckled dark brown above, paler below, pale
inner primaries in flight. Adult plumage after 3 years.

HABITAT Year-round resident or winter visitor almost throughout
Europe. Breeds colonially on all types of coast, moorland and town
buildings. Occurs in almost any habitat. Favours refuse dumps.

NEST Bulky mound of grass and flotsam on ground.

VOICE Noisy, familar laughing *kay-ow* or *yah-yah-yah* and mewing
calls.

GENERAL Widespread, often very numerous.

adult
summer

ADULT The largest European gull (68 cm). Note massive beak, jet black back and pink legs. In flight shows all-black wings with white trailing edge and white spots at tips of primaries. Sexes similar.

JUVENILE Speckled brown above, paler below. Note dark trailing edge and paler inner primaries in flight. Adult plumage after 4 years.

HABITAT The most maritime gull. Breeds in N and W on islands and cliffs. Year-round resident in most areas, some disperse far out to sea.

NEST Seaweed and flotsam nest on ledge.

VOICE Gruff, powerful *kow-kow-kow*.

GENERAL Widespread, not as numerous as other gulls. Rare Glaucous Gull *(L. hyperboreus)* adult has white wingtips, immature uniformly pale brown.

Common Gull

Larus canus

ADULT Medium (40 cm), grey-backed gull with rounded head and smallish yellow beak. Watch for greenish yellow legs and, in flight, black wingtips with white spots. In winter, has heavy grey flecking on head. Sexes similar.

JUVENILE Has speckled head, grey back, blackish wings and black-tipped white tail.

HABITAT Summer visitor breeding on N coasts, hillsides and moors. Year-round resident, migrant or winter visitor in S. Primarily coastal, but often on grassland and fields on migration.

NEST Grassy cup on ground.

VOICE High-pitched *key-yaa* and nasal *gah-gah-gah*.

GENERAL Widespread in many habitats, often fairly common.

adult
summer

Rissa tridactyla

ADULT Medium (40 cm), slender-winged gull. Watch for crimson-lined yellow beak, black legs. Wings long and slim, held angled in buoyant flight, grey with black tips and no white spots. Sexes similar.

JUVENILE As adult, but with black spot behind eye, black collar mark, black-tipped, slightly forked white tail and bold, black M on upper surface of wings.

adult summer

HABITAT A maritime gull, many disperse widely across Atlantic in winter. Nests colonially on cliff ledges, occasionally on buildings on W and N coasts.

NEST Guano, mud and seaweed glued to sheer cliff, often under overhang.

VOICE Noisy at colonies, diagnostic *kitti-wake*.

GENERAL Widespread, locally numerous.

Little Gull

Larus minutus

**adult
summer**

ADULT Medium (27 cm), but small and dainty for a gull. Watch for jet black cap, short dark red bill and legs. In flight shows pale grey upperside to wings, note dark grey underside and rounded white tips. In winter, loses black hood, has dark spot behind eye. Sexes similar.

JUVENILE As winter adult, with striking black M across wings.

HABITAT Summer visitor breeding on Baltic marshes, year-round resident, migrant, or winter visitor to W and S coasts. Most winter at sea, occurs on large inland fresh waters on migration.

NEST Grassy mound on swampy ground.

VOICE High-pitched *kar-eee* and *kek-kek-kek*.

GENERAL Regular, locally fairly common, increasing. Note dipping feeding flight.

**adult
summer**

ADULT Medium (40 cm), large for a tern, and heavy in flight. Watch for long yellow-tipped black beak, bristling black crest. In flight, grey wings show darker primaries; rump white, tail white and slightly forked. Forehead white in autumn. Sexes similar.

JUVENILE As autumn adult, with more white on crown and scaly blackish markings on back.

HABITAT Summer visitor or migrant along coasts. Breeds on isolated beaches and islands.

NEST Simple scrape in sand.

VOICE Distinctive and loud *kay-reck* or *kirr-ick*.

GENERAL Widespread coastally, erratic but locally common as breeding bird. Gull-billed Tern (*Gelochelidon nilotica*): similar, shorter black beak and grey rump. Often occurs inland.

adult

Common Tern

Sterna hirundo

ADULT Medium (35 cm) sea tern. Note deeply forked tail with long streamers. Watch for black cap and black-tipped red beak. Wings uniformly pale grey, primaries with blackish border. Forehead white in autumn. Sexes similar.

JUVENILE As autumn adult, with brown markings across back and black forewing edge.

HABITAT Migrant summer visitor, breeding coastally and occasionally inland over much of Europe. Nests colonially on beaches and on coastal, estuarine and freshwater islands.

NEST Scrape in grass or sand.

VOICE Swift *kirri-kirri-kirri*; harsh *kee-aarh* with emphasis on second syllable.

GENERAL Widespread and fairly common, often numerous around colonies.

adult

Sterna paradisaea

adult

ADULT Medium (37 cm) sea tern, shorter-legged and more grey-bellied than Common Tern (p.124). All-red beak. In flight, watch for forked tail with long streamers, and translucent primaries. Forehead white in autumn. Sexes similar.

JUVENILE As autumn adult, with grey-brown markings across back and grey forewing edges contrasting with white trailing edge.

HABITAT Migrant along W coasts, breeding colonially on N islands. Unusual inland.

NEST Shallow scrape in sand or grass.

VOICE Short sharp **kee-aah**.

GENERAL Widespread, locally common. Rare Roseate Tern (*S. dougallii*) very pale, with extremely long tail streamers, almost all-black beak.

ADULT Small (22 cm), stubby sea tern with distinctive flicking flight. Watch for shallowly forked tail, black-tipped yellow beak, yellow legs and white forehead patch. In flight shows conspicuously black outer primaries. Sexes similar.

JUVENILE Scaly-backed version of autumn adult.

HABITAT Migrant along W and S coasts and lagoons, breeds in loose colonies on sandy beaches.

NEST Shallow scrape in sand.

VOICE High-pitched kitick, hurried *kirri-kirri-kirrick*.

GENERAL Widespread and regular, but nowhere numerous.

adult

Chlidonias niger

ADULT Small (25 cm), unmistakable when breeding, grey-winged, sooty black-bodied marsh tern. Note white undertail coverts. Watch for black beak and legs and shallowly forked tail. In flight, dips to pick food off water. In autumn and winter, white body, nape and forehead, black crown and dark vertical half-collar marks. Sexes similar.

adult

JUVENILE As winter adult, but scaly above.

HABITAT Migrant and summer visitor to S, central and E Europe, breeds colonially on freshwater swamps and marshes. Migrates along coasts and over inland waters.

NEST Semi-floating platform of waterweed.

VOICE Rarely heard *krit* or *kreek*.

GENERAL Reasonably widespread, regular, locally fairly common.

adult

ADULT Medium (40 cm) auk, upright on land, swims low in the sea, diving frequently. Watch for dagger-like beak, chocolate brown upperparts. In winter, sooty black above, white throat and face. Sexes similar.

JUVENILE Resembles winter adult.

HABITAT At sea for much of year off N and W Europe; breeds colonially on cliff ledges.

NEST Single egg laid on bare open rock ledge.

VOICE Grumbling growls and croons on ledges, silent elsewhere.

GENERAL Locally common along rocky breeding coasts in summer, scarce and erratic elsewhere. Black Guillemot (*Cepphus grylle*) of NW coasts and adjacent seas is all-black with striking white wing patches in summer; grey backed, white elsewhere in winter. Note vermillion legs.

Alca torda

ADULT Medium
(40 cm), squat, thick-necked
auk, jet black above, white below.
Dives frequently. Watch for deep flat
beak with white vertical line and fine
white stripe leading from beak to eye.
Winter birds duller, with white face and throat.
Flight whirring low over sea. Sexes similar.

JUVENILE As winter adult, greyer with slimmer beak.

HABITAT Summer visitor breeding in loose colonies on
W and N cliffs, winters at sea.

NEST Single egg laid on bare rock in cavity.

VOICE Low growls.

GENERAL Locally common along rocky coasts
in summer, scarce and erratic elsewhere.
Arctic-breeding Little Auk *(Plautus alle)*:
half the size of Razorbill, black above
and on throat, white on belly, small
triangular beak. Irregular in winter.

adult
summer

ADULT Medium (30 cm), squat, upright and familiar auk. Watch for black back, white belly and face patch, bright orange legs and webbed feet, and colourful parrot-like beak. Winter birds duller, with grey face and smaller, dark beak. Sexes similar.

JUVENILE Greyer version of winter adult with slimmer dark beak.

HABITAT Summer visitor breeding colonially on remote headlands and islands in W and N Europe, winters out at sea.

NEST Single egg laid down burrow.

VOICE Low growls.

GENERAL Locally common, sometimes numerous, on breeding coasts in summer. Scarce and erratic elsewhere.

adult

adult female

ADULT Medium (33 cm), slim, short-legged and long-tailed, heard more than seen. Watch for grey body, barred underparts and white-tipped black tail. Beak tiny; legs yellow. Falcon-like flight on fluttering curved wings; tail looks spoon-ended. Sexes normally similar, female rarely chestnut.

JUVENILE Mottled dark brown above, white with blackish barring below.

HABITAT Migrant and summer visitor to all of Europe except far N, occurs in woodland, on heaths, moors, marshes and farmland.

NEST Parasitic, lays eggs in foster parent nests.

VOICE Cuck-oo and variants, throaty chuckle; female uses bubbling trill.

GENERAL Widespread, often fairly common.

ADULT Medium (28 cm), slim, fast-flying pigeon. Watch for pink breast, scaly bronze back and shoulders, black and white-barred collar patches. In flight shows bronze wings with grey diagonal bars, longish black tail with narrow white borders. Sexes similar.

JUVENILE Duller, browner version of adult, lacking collar marks.

HABITAT Migrant and summer visitor, breeding over much of Europe except N. Essentially a woodland, scrub and farmland bird.

NEST Flimsy platform of twigs in bush.

VOICE Far-carrying monotonous and prolonged purring.

adult

GENERAL Widespread, locally fairly common.

Columba palumbus

ADULT Medium (40 cm), cumbersome pigeon. Watch for pink breast, white collar marks and diagnostic white bar visible in closed wing, conspicuous in flight. Flight fast, but noisy and clumsy, often colliding with vegetation. Gregarious. Sexes similar.

JUVENILE As adult, but lacks collar marks.

HABITAT Widespread year-round resident, summer visitor to N and NE Europe. Breeds in woodland and scrub, feeds in woodland, on all farmland and in urban areas.

adult

NEST Flimsy platform of twigs in bush or tree.

VOICE Monotonously repetitive **coo-coo, coo-coo**.

GENERAL Widespread, often common, often in flocks. Can damage crops.

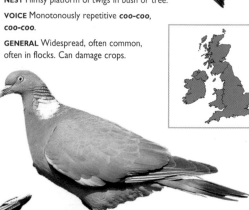

Stock Dove

Columba oenas

ADULT Medium (33 cm), dull-grey pigeon. Watch for pinkish flush on breast and metallic green collar marks. Swift, direct flight showing triangular black-bordered grey wings and dark-tipped grey tail. Sexes similar.

JUVENILE As adult, but duller, lacking collar marks.

adult

HABITAT Year-round resident over much of Europe, summer visitor to N and NE. Mainly on farmland and woodland, occasionally on coasts and marshes.

NEST In hollow tree or burrow.

VOICE Booming *coo-oo* or *coo-roo-oo*.

GENERAL Widespread, locally fairly common. Rock Dove (*C. livia*, ancestor of town and racing pigeons): similar, with double black wingbars and conspicuous white rump. On remote rocky N coasts and S mountains; scarce.

adult

ADULT Medium (30 cm), sandy pigeon. Pinkish head and neck with white-edged black band round nape. In flight looks long-tailed and hawk-like, shows buff and grey wings with blackish primaries. Long tail buff above, black below, showing much white on underside. Gregarious. Sexes similar.

JUVENILE Dull version of adult, lacking collar mark.

HABITAT Westward-spreading, post-1940 newcomer to Europe from Asia, colonising Ireland in 1970s. Year-round resident of Europe, except N, in farmland, parks and towns.

NEST Flimsy twig platform in bush, tree, ledge.

VOICE Distinctive dry *aaah* in flight; song strident and persistent *coo-coo-coo*.

GENERAL Widespread, comparatively common.

ADULT Medium (35 cm), pale, upright owl. Watch for finely mottled orange-buff upperparts and heart-shaped white facial disc with large dark eyes. Legs long, knock-kneed and feathered. Underparts white in NW Europe, dark-speckled rich buff elsewhere. Long-winged in flight, legs dangling.

JUVENILE Much as adult.

HABITAT Widespread year-round resident except in N and NE Europe. Favours open woodland, farmland, heath and marshes.

adult

NEST In hollow tree or deserted building.

VOICE Usually quiet; snoring noises near nest, occasional strident shriek elsewhere.

GENERAL Widespread, nowhere numerous. Usually nocturnal, but in winter may hunt in daylight.

Surnia ulula

ADULT Medium (38 cm), long-tailed owl, active daylight hunter. Watch for speckled back and finely barred underparts, and for rectangular white facial disc with striking black vertical margins. Hawk-like posture emphasized by long tail and short, rounded wings in flight. Sexes similar.

JUVENILE Much as adult.

HABITAT Year-round resident in N birch and conifer forests.

NEST In tree hole.

VOICE Distinctive rapid series of short whistles.

GENERAL Helpfully chooses prominent perches. Locally fairly common.

DISTRIBUTION
N Scandinavia and Finland, not found in Iceland and Greenland

adult

ADULT Small (23 cm), squat, upright owl, perches prominently in daylight. Watch for rectangular facial disc with white 'spectacles', and 'fierce' white eyebrows. Sexes similar.

JUVENILE As adult, but paler, heavily streaked.

HABITAT Year-round resident over much of Europe except N and NW. Broad choice of habitats from woodland, farm and heath to suburban areas and coasts.

adult

NEST In hollow in building, bank or tree.

VOICE Penetrating yelps and **poop** whistles.

GENERAL Widespread, fairly common. Pygmy Owl *(Glaucidium passerinum)*: comparatively tiny, with rounded head and pale facial disc. Hunts in daylight through N conifer forests.

Tawny Owl

Strix aluco

ADULT Medium (38 cm), plump and distinctively round-headed owl. Plumage finely marked grey-brown to reddish brown. Watch for circular facial disc, narrowly bordered black and buff, with two central prominent buff stripes up onto crown. Large all-dark eyes. Sexes similar.

adult

JUVENILE Much as adult.

HABITAT Year-round resident over much of Europe except N and Ireland. Broad choice of habitats from woodland and farmland to urban areas with large trees.

NEST Usually in hollow tree.

VOICE Well-known trembling *whoo-hoo-hoooo* and sharp *kew-wit*.

GENERAL Widespread and familiar despite nocturnal habits. The most numerous owl.

adult

ADULT Medium (35 cm), slim and upright owl, comparatively long-winged in flight. Plumage finely marked rich browns giving excellent camouflage. Watch for rounded head, circular facial disc with buff lateral margins and paired white central stripes. Conspicuous long ear-tufts. Eyes strikingly yellow or flame. Sexes similar.

JUVENILE As adult, but duller.

HABITAT Year-round resident or migrant over much of Europe, summer visitor in N. Favours woodlands of many types.

NEST Often in old crow's nest or squirrel drey.

VOICE Repetitive deep ***poop*** calls when breeding.

GENERAL Nocturnal, widespread, but inconspicuous, even secretive. Probably commoner than seems.

Asio flammeus

adult

ADULT Medium (38 cm), pale sandy-brown, daylight-hunting owl. Watch for short indistinct ear-tufts; clear roughly circular facial disc, bright yellow eyes. Distinctive bouncing flight; note pale undersides to wings and conspicuous dark patches at wrist. Sexes similar.

JUVENILE Much as adult.

HABITAT Year-round resident or migrant in central and W Europe. Summer visitor in N and winter visitor to S. Favours tundra, moor, rough grassland and marshes.

NEST Rough scrape on ground.

VOICE Normally silent.

GENERAL Widespread, but erratic in distribution and numbers, sometimes locally numerous..

Tengmalm's Owl

Aegolius funereus .145

DISTRIBUTION
N Scandinavia
and Finland, not
found in Iceland
and Greenland

ADULT Small
(25 cm), squat,
large-headed owl.
Brown above,
boldly spotted;
whitish below with brown mottling and streaking. Watch
for rectangular facial disc with pale then dark borders, and for
conspicuous 'raised eyebrow' appearance. Eyes yellow. Sexes
similar.

JUVENILE Much as adult.

HABITAT Year-round resident in N and E.
Favours woodland and forest, often montane
and often predominantly coniferous.

adult

NEST Usually hollow trees or old nests.

VOICE Repetitive abrupt whistle.

GENERAL Largely nocturnal; variable in both
distribution and numbers, as in several other owls depending
on availability of suitable prey.

Caprimulgus europaeus

ADULT Medium (28 cm), slim and exceptionally well camouflaged finely mottled and streaked brown, buff and grey plumage. Note short-legged horizontal stance. In silent, moth-like flight, watch for long tail and pointed wings. Males show white patches in wingtips and at tip of tail.

JUVENILE Much as adult female.

HABITAT Summer visitor or migrant to much of Europe except far N. Favours dry heaths, open (often recently cleared) woodland and scrub.

NEST Well-concealed simple scrape on ground.

VOICE Listen for extended *churring*, with wing-claps in display flight. Best heard at dusk.

GENERAL Widespread, but only locally regular, rarely numerous.

adult male

adult male

ADULT Small (18 cm), highly aerial bird with familiar long, slim sickle-shaped wings. Note solid, but well-streamlined, sooty-black body, short shallowly forked tail, and large head with smoky white throat patch. Often gregarious, flying at high speed in noisy groups. Sexes similar.

JUVENILE As adult, but with scaly markings.

HABITAT Summer visitor or migrant over most of Europe except far N. Usually breeds in urban areas, feeds over any habitat, often over fresh water.

NEST Rough crudely lined scrape in roof cavity.

VOICE Distinctive shrill high-pitched scream.

GENERAL Widespread, often common. Pallid Swift *(A. pallida)* of Mediterranean: similar but stockier, slower in flight, slightly paler with larger throat patch.

Apus melba

DISTRIBUTION
Summer visitor
to S Europe and
N Africa; winters
in S Africa

ADULT Small (20 cm), but detectably larger than Swift (p.147). Shares Swift flight silhouette and long, narrow sickle-shaped wings, but watch for sandy brown upperparts, throat and undertail contrasting with white belly. Flight more powerful and faster even than Swift. Sexes similar.

JUVENILE As adult, but duller with scaly back markings.

HABITAT Summer visitor to S Europe, breeding in mountain areas, towns and on coastal cliffs. Ranges widely when feeding.

NEST Cavity in rocks or building; breeds colonially.

VOICE Surprisingly loud, far-carrying and distinctive musical trill.

GENERAL Fairly widespread, locally common.

adult
male

ADULT Small (20 cm including tail streamers) and familiar. Watch for dark purplish upperparts, white underparts and chestnut face patch. Swift, swooping flight on long curved wings. Shows white in deeply forked tail in flight. Male has longer streamers, otherwise sexes similar.

JUVENILE Duller, with short tail streamers.

adult male

HABITAT Summer visitor to most of Europe. Breeds in buildings, ranges widely when feeding, often over water.

NEST Open cup of mud and grass.

VOICE Extended musical twittering; sharp chirrup of alarm indicates presence of a raptor.

GENERAL Widespread. Scarcer Red-rumped Swallow (*H. daurica*), of Mediterranean: blackish cap, chestnut cheeks, nape and rump contrasting with dark back.

Riparia riparia

ADULT Tiny (12 cm) hirundine. Largely aerial; watch for longish curved, pointed wings. Sandy brown above, whitish below with brown collar. Gregarious. Sexes similar.

adult male

JUVENILE As adult, but sandy scaly markings on back.

HABITAT Summer visitor and migrant to all Europe except farthest N. Breeds colonially in sandy banks, usually feeds in flight over nearby fresh waters.

NEST Excavates burrow in bank.

VOICE Soft rattling trill, sharp chirrup of alarm.

GENERAL Widespread, locally numerous. Has declined dramatically in some areas recently. Similar Crag Martin (*Hirundo rupestris*): grey-brown above, grey-buff below. Heavier-built and broader-winged than Sand Martin, year-round resident in Mediterranean mountains and occasionally towns.

House Martin

Delichon urbica

ADULT Tiny (12 cm) hirundine with distinctive purplish black and white plumage. In flight, watch for relatively short broad-based curved wings, white rump and short shallowly forked tail. From beneath, black cap contrasting with white undersides gives capped appearance. On ground, watch for white legs feathered to toes. Sexes similar.

JUVENILE Much as adult, but duller.

HABITAT Summer visitor or migrant over most of Europe except furthest N. Breeds on buildings, occasionally cliffs. Ranges widely when feeding, often over water.

NEST Very distinctive quarter-sphere of mud pellets, fixed under overhang. Often colonial.

VOICE Harsh chirrup; unmusical twittering.

GENERAL Widespread, often fairly numerous.

adult
male

Kingfisher

ADULT Small (17 cm), but unmistakable.
Upperparts electric blue-green,
underparts chestnut. Crown blue, cheek
stripe chestnut and white. Arrow-like rapid flight,
usually low over water. Tiny scarlet-orange feet
and large dagger-shaped black or black and orange
beak. Sexes broadly similar.

adult

JUVENILE As adult, but duller, with heavily dark-flecked crown.

HABITAT Year-round resident over much of Europe, migrant or
summer visitor in N. Favours rivers, lakes and streams;
occasionally coasts in winter.

NEST Excavates burrow and nest
chamber in earth bank beside water.

VOICE Distinctive shrill **tseet** or
chee-tee.

GENERAL
Widespread, but
nowhere numerous.

Bee-eater

Merops apiaster

DISTRIBUTION
Summer visitor to
Spain, S. France,
Italy, Greece,
Turkey

ADULT Medium
(28 cm), slim,
swallow-like and
unmistakably
colourful. No
other European bird shows such dazzling plumage. Watch for long-
winged, swooping flight and slim, extended, central tail feathers,
longish, dark, downcurved and pointed beak. Gregarious. Sexes
similar.

JUVENILE Muted-colour version of adult.

HABITAT Favours open dry country. Often
feeds over lakes and marshes with high
insect populations.

adult

NEST Colonial, excavates burrow in sandy soil
or banks.

VOICE Listen for distinctive bell-like trilling *prrewit*,
often audible when birds are out of sight.

GENERAL Only locally common.

Hoopoe

154 *Upupa epops*

ADULT Medium (28 cm) and unmistakable. Watch for black and white striped back and wings, unusual pinkish-fawn body, long black and ginger crest (erected when excited or often on landing), long, slender, slightly downcurved beak. Distinctive floppy flight, black and white pattern prominent on rounded, fingered wings. Sexes similar.

JUVENILE As adult, duller and greyer, with tiny crest.

HABITAT Summer visitor or migrant to much of Europe except N and NW, where occasional vagrant. Favours dry open country with trees, including orchards, cork oak and olive groves.

NEST Notoriously smelly; in tree hole.

VOICE Soft, but penetrating, repeated *poo*.

GENERAL Widespread, but only locally common.

adult

adult

ADULT Small (18 cm) relative of woodpeckers. Short-legged, with relatively long body. Looks drab brown at a distance, but close to note beautiful finely marked plumage. Watch for striped head and grey and buff V markings on back. Beak short and strong. Tail long, soft (not stiff as in woodpeckers), finely barred. Often feeds on ground. Sexes similar.

JUVENILE As adult, but duller.

HABITAT Summer visitor or migrant to most of Europe except far N and NW. Favours open land with old trees.

NEST Excavates hole in tree.

VOICE Persistent laughing ***kee-kee-kee***.

GENERAL Widespread, but erratic and inconspicuous, never numerous.

juvenile

ADULT Medium (30 cm) woodpecker, often feeds on ground. Distinctive greenish-gold upperparts, gold rump, greenish buff below. Watch for stout dagger-like beak, red crown and black face. Male has red and black moustachial stripe, female has black stripe. Flight undulating.

JUVENILE Duller, with dense darker barring.

HABITAT Year-round resident over much of Europe except N and Ireland. Favours dry heath, grassland and open woodlands.

NEST Makes oval-opening hole in tree.

VOICE Distinctive ringing laugh *yah-yah-yah*.

GENERAL Widespread, locally fairly common. Grey-headed Woodpecker (*P. canus*) of E Europe: browner, grey head and small scarlet patch on crown of male only.

Black Woodpecker

Sterna albifrons

DISTRIBUTION
C, E and N Europe.
Population growing
in NW Europe

ADULT Medium
(45 cm), but the
largest and most
striking of
European
woodpeckers. Plumage almost entirely glossy black, with
crimson crown, more extensive in male than female. Golden eye.
Watch for long-necked, long-tailed 'stretched' appearance in
undulating flight.

JUVENILE Only slightly duller than adult, eye pale.

HABITAT Primarily a bird of E Europe, but
does occur in the Pyrenees. Favours
extensive areas of old forest of all types.

NEST Excavates hole in tree.

VOICE Harsh, far-carrying *klee-oh*. Drums
frequently, loud and slow rhythm.

GENERAL Fairly widespread and locally
not uncommon.

juvenile

ADULT Small (23 cm), pied woodpecker. Watch for large white shoulder patch, multiple white wingbars, conspicuous scarlet undertail. Complex head pattern; crown black in female, with red nape patch in male. Undulating flight, perches head-up on trees.

JUVENILE Duller version of adult, note red crown.

HABITAT Widespread year-round across much of Europe except Ireland and farthest N regions.

NEST Excavates hole in tree.

VOICE Explosive *chack*; drums frequently.

GENERAL Widespread, often common. Middle Spotted Woodpecker *(D. medius)* of central Europe is smaller, with all-red crown and dull pink undertail, streaked buff underparts.

juvenile

Lesser Spotted Woodpecker

ADULT Small (15 cm), sparrow-sized woodpecker. Watch for black and white 'ladder' markings on back and wings. Male has white forehead and red crown, female buffish white.

JUVENILE Much as adult, but with reddish crown.

HABITAT Year-round resident over much of Europe except extreme N and NW, including Ireland. Favours deciduous woodland, parks, orchards.

juvenile

NEST Excavates hole in tree.

VOICE Usefully distinctive high-pitched, repetitive **kee-kee-kee**. Extended high-pitched drumming frequent when breeding.

GENERAL Widespread, but often inconspicuous.

Galerida cristata

DISTRIBUTION
Found throughout
Europe apart from
far north and west

adult

ADULT Small (17 cm), well-camouflaged, buffish lark. Long crest almost always erect and visible. In flight, watch for sandy-brown tail with distinctive chestnut outer feathers. Spends much time on the ground, running swiftly. Sexes similar.

JUVENILE As adult, but often more rufous, with smaller crest.

HABITAT Year-round resident across S and central Europe, absent from N and (unexpectedly) from Britain and Ireland. Favours open land, frequently farmland and roadsides, often near habitation.

NEST Well-concealed grassy cup on ground.

VOICE *Doo-dee-doo*; varied melodious song with mimicry, usually from ground or a post.

GENERAL Widespread, frequently common.

Woodlark

Lullula arborea

ADULT Small (15 cm), stockily-built, short-tailed lark. Watch for rich brown appearance and bold whitish eyestripes, which with chestnut cheek patches give capped appearance. In flight shows black and white patch on wing shoulder and white tips to outer tail feathers. Sexes similar.

JUVENILE Much as adult.

HABITAT Resident or migrant in SW and S Europe, less common as summer visitor to central and W areas. Favours dry open woodland and heaths.

adult

NEST Well-concealed grassy cup on ground.

VOICE Listen for distinctive flight call **tee-loo-ee**; melodious song (in spiralling song flight) based on **loo-loo-yaa** phrases.

GENERAL Widespread, locally fairly common.

Skylark

Alanda arvensis

adult

ADULT Small (18 cm), long-bodied, well-camouflaged lark. Watch for short but often visible crest. In flight shows white outer tail feathers and characteristic white trailing edges to markedly triangular wings. Sexes similar.

JUVENILE Much as adult.

HABITAT Year-round resident and migrant in S, central and W Europe, summer visitor in N. Frequents open landscapes of all types.

NEST Well-concealed grassy cup on ground.

VOICE Flight call a liquid chirrup; varied musical song rich in mimicry, usually while hovering or circling high.

GENERAL Widespread, often common. Rare Shore Lark (*Eremophila alpestris*) breeds on Arctic tundra, winters on remote coastal marshes, has yellow and black face and bib.

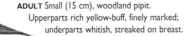

ADULT Small (15 cm), woodland pipit. Upperparts rich yellow-buff, finely marked; underparts whitish, streaked on breast. Watch for pale pinkish legs, white outer tail feathers in flight. Sexes similar.

juvenile

JUVENILE Much as adult.

HABITAT Summer visitor or migrant to most of Europe. Favours heaths with trees and woodland with substantial clearings.

NEST Well-concealed grassy cup on ground.

VOICE Distinctive **teees** flight call. Descending trilling song, ending in repeated **see-ar** notes, in parachute song flight.

GENERAL Widespread, locally fairly common.

ADULT Small (15 cm), undistinguished, streaky pipit, largely terrestrial in behaviour. Plumage variable from yellowish, through olive to greenish-buff or brown, whitish below, copiously streaked. Watch for pale brown legs, white outer tail feathers. Sexes similar.

JUVENILE Much as adult.

HABITAT Year-round resident, winter visitor or migrant over much of Europe, summer visitor to N. Favours open landscapes: moorland, heath, grassland, farmland and marshes.

NEST Well-concealed grassy cup on ground.

VOICE Flight call a thin *tisseep* or *tseep;* song an accelerating descending trill, weaker than Tree Pipit (p.159).

GENERAL Widespread, locally common.

juvenile

ADULT Small (17 cm), pale pipit. Upperparts pale sandy-buff, only faintly marked, underparts whitish, flushed with pink in spring. Watch for conspicuous pale eyestripe, and pinkish legs. Relatively long-tailed, behaves almost more like a wagtail than a pipit. Sexes similar.

JUVENILE Rather darker and more heavily streaked.

HABITAT Summer visitor to S and central Europe, vagrant further N. Favours arid open areas: heaths, saltpans, dunes and marshland.

NEST Well-concealed grassy cup on ground.

juvenile

VOICE Characteristic broad *tseep* flight call; song repetitive, reeling *seely-seely-seely*.

GENERAL Widespread, locally fairly common.

DISTRIBUTION
Summer visitor to C and S Europe; not seen in Britain, Iceland and Scandinavia

ADULT Small (17 cm) pipit, darker and greyer overall than Meadow Pipit (p.164), longer in the tail. Watch for longish dark legs and smoky grey outer tail feathers. Sexes similar.

JUVENILE Much as adult.

HABITAT Year-round resident, winter visitor or migrant along much of W coast of Europe, summer visitor to Scandinavian coasts. Favours rocky coasts.

NEST Well-concealed grassy cup.

VOICE Strident **zeep**; loud descending trill song in parachute display flight.

GENERAL Widespread. Scarcer Water Pipit (*A. spinoletta*): paler, with unstreaked back, unstreaked pink breast in spring (whitish, boldly streaked at other times); breeds in S and E mountainous areas, winters in S marshlands, vagrant elsewhere.

juvenile

adult

ADULT Small (17 cm), short-tailed wagtail. Males have yellow underparts, white-edged black tails; heads vary. Blue-headed (W, central): blue head, white eyestripe; Yellow (NW, Britain and Ireland): olive head, yellow eyestripe; Spanish (Iberia): grey head, white bib, white behind eye; Grey-headed (N): dark grey head, black cheeks, no eyestripe; Black-headed (SE): jet black head, no eyestripe. All females olive above, dull yellow below. In winter, duller and paler.

JUVENILE As winter adult, with scaly wing markings.

HABITAT Widespread summer visitor and migrant. Favours open land: farmland, marshes, grassland.

NEST Well-concealed grassy cup on ground.

VOICE *Tseep* flight call; twittering song.

GENERAL Locally fairly common.

adult male summer

adult female summer

ADULT Small (18 cm). Slimmest and longest-tailed of European wagtails. Watch for grey back and crown, white eyestripe, yellow underparts, brilliant yellow rump and undertail. Wags white-edged, blackish tail non-stop. Male is brighter yellow, has black bib in summer.

JUVENILE Paler, duller version of female.

HABITAT Year-round resident over much of Europe, summer visitor to N and NE, and to some mountain areas. Favours fast-moving fresh water, streams, rapids, weirs and sluices.

NEST Grassy cup hidden in cavity near water.

VOICE Characteristic *chee-seek* call; trilling song resembles Blue Tit (p.218).

GENERAL Widespread, never numerous.

adult male summer

ADULT Small (18 cm), pied wagtail with silver-grey back and incessantly wagging, white-edged black tail. Grey crown, white cheeks, black bib. Female duller and less clearly marked than male. Often gregarious. Undulating flight.

adult female winter Pied

adult male summer Pied

adult female summer Pied

JUVENILE As female, but with smoky-yellow tinge.

HABITAT Year-round resident or migrant over much of Europe, summer visitor in N. Favours open grassland, farmland, marshland and waterside, and urban areas.

NEST Grassy cup concealed in cavity.

VOICE Soft disyllabic *swee-eep*; twittering song.

DISTRIBUTION Resident in W and S Europe; summer visitor in Scandinavia and E Europe

GENERAL Widespread, locally common. Pied Wagtail of Britain and Ireland is dark subspecies, male with jet black crown and back. Sharp *chissick* call.

Bombycilla garrulus

juvenile

ADULT Small (17 cm), plumply Starling-like shape and flight. Watch for pinkish-brown plumage, black bib and face, drooping crest. In flight shows yellow tip to blackish tail. Red 'waxy' ends to wing feathers visible only at close range. Sexes similar.

JUVENILE As adult, but duller, lacking red feather tips.

HABITAT Year-round resident or winter visitor to N Europe, summer visitor to far N. Breeds in conifer woodland or boreal scrub, winters where berries plentiful.

NEST Grassy cup in tree fork.

VOICE Characteristic bell-like trill.

GENERAL Locally common in breeding areas. Erratic wanderer elsewhere.

ADULT Small (17 cm) and dumpy, like a gigantic aquatic Wren (p.171). Bobs, tail-cocked, on rocks before walking into fast moving water. Large white bib, belly chestnut (Britain and Ireland) or blackish (rest of Europe). Sexes similar.

JUVENILE Duller and scaly.

HABITAT Year-round resident in N, NW and S Europe. Favours fast-moving rivers and streams, often in hilly or mountainous country. Vagrant elsewhere.

NEST Grassy cup, concealed in cavity or under overhang, always beside or over water.

VOICE Loud, distinctive *zit* or *zit-zit* call; both sexes produce warbling song.

GENERAL Fairly common in appropriate habitat.

adult
British race

Wren

Troglodytes troglodytes

juvenile

ADULT Tiny (10 cm) but familiar despite its mouse-like, largely terrestrial habits. Watch for crouched-stance, dark-barred rich brown plumage, pale eyestripe, cocked tail and pointed, downcurved beak. Flight whirring on rounded wings, usually low and direct. Sexes similar.

JUVENILE Much as adult.

HABITAT Year-round resident over much of Europe, summer visitor in far N. Favours dense vegetation, also rocky mountains and sea cliffs.

NEST Domed grassy structure with side entrance, well-concealed in vegetation or cavity.

VOICE Scolding *churr* call; amazingly loud boisterously musical song.

GENERAL Widespread, often common.

Dunnock

Prunella modularis 173

juvenile

ADULT Small (15 cm), dull bird, lead-grey on head and breast, dark brown back and wings. Beak straight and pointed, legs pinkish, strong; spends much time hopping on ground or in vegetation. Sexes similar.

JUVENILE Duller and scaly.

HABITAT Year-round resident over much of Europe, summer visitor to far N and NE, winter visitor to far S. Favours woodland and scrub of all types, farmland and urban gardens.

NEST Well-concealed grass cup in shrub.

VOICE Strident piping *seek* call; brief but melodious snatches of song.

GENERAL Widespread, locally fairly common. Scarce Alpine Accentor *(P.collaris)*: of high mountain areas in S Europe, broadly similar, but with speckled grey bib and chestnut breast and flanks.

Luscinia megarhynchos

juvenile

ADULT Small (17 cm), drab thrush. Watch for warm brown back, paler underside, relatively long rufous tail. Long, strong, pinkish-brown legs. Keeps under cover. Sexes similar.

JUVENILE Paler, heavily speckled.

HABITAT Summer visitor or migrant to S and central Europe. Prefers woodland with dense undergrowth, scrub, and swampy thickets.

NEST Well-concealed leafy cup near ground.

VOICE Fluid **hoo-eet** call; song rich and varied, long and melodious; listen for opening pee-ooo notes.

GENERAL Widespread, locally quite common. Similar Thrush Nightingale *(L. luscinia)* in same habitats further N, greyish below, faint speckling.

ADULT Small (15 cm), often secretive thrush. Male has electric blue throat with red or white central spot, much duller in autumn and winter. Female has black-fringed white throat. Watch for brown tail, darker at tip, chestnut patches on either side at base – often all that is seen as darts for cover.

adult

JUVENILE Sandy brown and speckled, with characteristic tail pattern.

HABITAT Summer visitor to N and NE Europe, migrant or vagrant elsewhere. Prefers dense, low, swampy scrub or heathland.

NEST Well-concealed grassy cup on ground.

VOICE Sharp *tack* call; extended high-pitched melodious warbling song.

GENERAL Locally fairly common.

adult

ADULT Tiny (13 cm), familiar, plump, long-legged thrush. Watch for rich orange-red face and breast with grey margin. Back brown, underparts whitish. Perky stance showing short brown tail, hops rapidly. Often terrestrial, flights usually short and low. Sexes similar.

JUVENILE Reddish-brown above with buff markings, whitish below heavily scaled with brown.

HABITAT Year-round resident, migrant or winter visitor over much of Europe, summer visitor to N and NE. Varied habitat from woodland, parks and gardens to offshore islands in winter.

NEST Well-concealed grassy cup on or near the ground, often in cavity.

VOICE Sharp *tick* call; high-pitched warbling song.

GENERAL Widespread, often common.

ADULT Small (15 cm), slim, red-tailed chat. Watch for grey back, white forehead, black face and chestnut underparts of summer male, colours partly concealed by buff markings at other times. Female brown above, pale buff below, but with characteristic brown-centred chestnut-red tail. Both sexes show plain brown wings in flight.

JUVENILE Speckled, with brown-centred red tail.

HABITAT Summer visitor or migrant across Europe. Favours woodland, parks and occasionally heaths.

NEST Well-concealed grassy cup, usually concealed in cavity.

VOICE Fluting *too-eet* call; brief, scratchy, but melodious song, ending in a rattle.

GENERAL Widespread, locally fairly common.

adult

Phoenicurus ochruros

juvenile

ADULT Small (15 cm), distinctively dark, red-tailed chat. Summer male sooty black, paler in winter. Watch for white wing patches. Female uniformly sooty buff. Both sexes have characteristic brown-centred chestnut-red tail. Often feeds on ground. Flicks and shivers tail.

JUVENILE As female, but heavily speckled buff.

HABITAT Year-round resident, migrant or winter visitor to W, SW and S Europe, summer visitor to N and E. Varied habitat including mountain screes, town roofs and major buildings.

NEST Grassy cup, well concealed in cavity.

VOICE Sharp *tack* call; brief rattling fast warbling song.

GENERAL Widespread, locally common in S.

ADULT Tiny (13 cm), upright chat. Watch for bold white eyestripe separating dark crown from equally dark cheeks, white moustachial streak, and orange-flushed breast of male. Female paler and duller. Chooses prominent perches, continuously flicks wings and tail. In flight shows white patches in wings and distinctive white sides to base of tail.

JUVENILE As female, but drabber, heavily speckled.

HABITAT Summer visitor or migrant over most of Europe. Favours open rough grassland, heath and scrub.

NEST Well-concealed grassy cup beneath bush.

VOICE Harsh *teck* call; brief high-pitched warble of song, usually produced in song flight.

GENERAL Though widespread, scarce in many areas.

adult
female

Saxicola torquata

ADULT Tiny (13 cm), plump, dark and upright chat. Watch for black head and striking white collar of male, dark brown head and indistinct paler collar patch in female. Chooses conspicuous perches, flicks wings and tail non-stop. In flight shows small white patch in wings and white rump.

JUVENILE As female, but drabber, heavily speckled.

HABITAT Year-round resident, migrant or winter visitor to S and W Europe, summer visitor to central and some N areas. Favours heath and scrub (often gorse).

NEST Grassy cup well-concealed on ground at base of bush.

VOICE Frequent *tchack* call; brief high-pitched scratchy warble song, often in song flight.

GENERAL Widespread, sometimes locally common.

ADULT Small (15 cm), pale, terrestrial chat. Watch for grey back, bold black eyepatch and wings of male; female browner and duller. Striking white rump and tail ending in an inverted black T mark conspicuous in flight. Fast bouncing hop across ground, flicks wings and tail frequently.

JUVENILE As female, but drabber, heavily speckled.

HABITAT Summer visitor or migrant to much of Europe. Favours open areas of heath, grass or moor, even coastal sand or shingle, rarely with tall vegetation.

NEST Grassy cup usually concealed in hole, old burrow, or crevice in the ground.

VOICE Harsh *tack*; brief scratchy warble of song, often produced in flight.

GENERAL Widespread, but only locally common.

adult

Black-eared Wheatear

Oenanthe hispanica **WHEATEARS**

ADULT Small (15 cm), terrestrial chat. Male upperparts whitish, washed cinnamon, underparts richer cinnamon, contrasting black wings. Black patch through eye, or black face and throat. Winter male much duller. Female as Wheatear (p.181), but darker head and wings. All have black T mark on white rump and tail, as Wheatear.

JUVENILE Similar to Wheatear, but darker-headed.

HABITAT Summer visitor or migrant to S and SW Europe. Favours open arid stony heath and scrub.

NEST Grassy cup in cavity, usually on ground.

VOICE *Tchack* call; brief high-pitched scratchy warbling song.

GENERAL Locally fairly common.

adult female

ADULT Small–medium (25 cm), but large among thrushes. Watch for dull plumage, sooty black in male, sooty brown with scaly markings in female. White crescentic throat patch clear in male, often obscure in female. In flight, look for silver-grey wings.

JUVENILE Rich brown, pale scaly markings, lacks bib.

HABITAT Summer visitor to N, W and central mountain areas, migrant almost anywhere. Favours upland grassland, moors, rocky mountainsides in breeding season.

NEST Well-concealed grassy cup on ground.

juvenile

VOICE *Chack* or *chack-chack* calls; song loud, simple, but melodious *chew-you, chew-you*.

GENERAL Though widespread, never numerous.

Blackbird

Turdus merula

ADULT Small–medium (25 cm), but large among thrushes. Adult male unmistakable in glossy jet black with orange beak and eye-ring. Female rich brown, with dark-bordered whitish throat, often faintly speckled on breast. Looks long-tailed in powerful direct flight.

JUVENILE Reddish brown above, slightly paler and spotted below.

juvenile

HABITAT Year-round resident, migrant and winter visitor over much of Europe, summer visitor to far N. Familiar in farmland, heath, woodland and urban areas.

NEST Grassy cup in tree or bush.

VOICE Penetrating *pink* or *chink* calls; extended fluting and melodious song. Chooses prominent song-posts.

GENERAL Widespread, frequently common.

DISTRIBUTION
Resident around
Mediterranean
along coasts;
widespread in
Spain

ADULT Small
(20 cm), dark
thrush. Summer
male unmistakable:
slate-blue body,
blackish wings. Winter male duller and slaty.
Female duller, brown above, fawn below with
darker streaks. Shy – creeps inconspicuously
around rocky areas.

adult female

JUVENILE Similar to female.

HABITAT Year-round resident in rocky, often mountainous areas
in S Europe, occasionally in towns.

NEST Grassy cup, usually concealed in crevice.

VOICE Sharp *tchick* call; loud musical song, often from prominent
rocky song-post.

GENERAL Widespread, but rarely numerous. Rock Thrush
(*M. saxatilis*): summer visitor to higher altitudes. Male blue above,
orange below with white back, black wings. Female brown above,
buff below.

juvenile

ADULT Small–medium (25 cm), long-tailed thrush. Adult has grey head, chestnut-bronze back, dark-speckled ginger-buff breast. Beak yellow, tipped black. Watch for black tail and grey rump in flight. Sexes similar.

JUVENILE Browner above, fawn below, heavily speckled.

HABITAT Winter visitor or migrant to much of Europe, year-round resident in north-central areas, summer visitor to far N. Breeds in woodland, forests, gardens. Winters in woodland, often on open farmland and grass.

NEST Grassy cup in tree fork.

VOICE Distinctive laughing *chack-chack-chack* calls; song a scratchy poorly-formed warble.

GENERAL Widespread, often quite common as breeding bird, migrant, and winter visitor.

juvenile

ADULT Small (23 cm), short-tailed, upright thrush. Familiar sandy-brown back, boldly black-speckled whitish underparts, tinged buff on breast. Medium-length and strength, pointed thrush beak. Often terrestrial, runs rather than hops. Flight direct, shows buff underwing. Sexes similar.

JUVENILE As adult, but heavily buff-speckled back.

HABITAT Year-round resident and migrant over central and S areas, winter visitor to SW Europe, summer visitor in N. Favours woodland, parks, gardens and other open landscapes with trees.

NEST Distinctive mud-lined grass cup in shrub.

VOICE Thin *seep* call; song usually a series of musical notes characteristically each repeated two or three times. Perches prominently to sing.

GENERAL Widespread, often common.

Redwing

188 *Turdus iliacus*

juvenile

ADULT Small (20 cm), dark, short-tailed thrush. Watch for buff eyestripe and moustachial streak on either side of dark cheek. Belly whitish, brown speckled, characteristic red flanks, red on underwings in flight. Sexes similar.

JUVENILE Duller than adult, heavily buff speckled on back.

HABITAT Summer visitor to N Europe, breeding in forests and gardens; migrant or winter visitor elsewhere. Favours woodland, open fields and grassland.

NEST Grassy cup in tree or shrub.

VOICE Extended *see-eep* flight call, especially migration. Song, fluting notes in slow tempo.

GENERAL Widespread, but erratic. Often locally numerous in winter.

ADULT Medium (27 cm), largest and palest of the European thrushes. Note pale sandy-brown upperparts and boldly brown-spotted pale buffish breast. In swooping flight, watch for whitish edges to tail. Sexes similar.

juvenile

JUVENILE Paler, greyer, grey scaly pattern on back.

HABITAT Year-round resident and occasional migrant over much of Europe, summer visitor to N and NE. Favours open woodland, farmland with trees, parks and gardens. Often on open fields in winter.

NEST Bulky and untidy cup of grass and litter, usually high in a tree.

VOICE Extended and angry-sounding churring rattle. Song melodious, simple and measured, often in early spring, uses prominent perches.

GENERAL Widespread, but rarely numerous.

ADULT Small (15 cm) warbler, unstreaked reddish-brown back and pale buffish underparts. Watch for buff eyestripe and characteristically longish rounded tail, often held fanned. Secretive, heard more than seen. Sexes similar.

JUVENILE Much as adult.

HABITAT Unusual among warblers in being year-round resident, occurring in damp, heavily vegetated marshes, ditches and scrub in S, SW and (erratically) W Europe.

NEST Well-concealed grass cup deep in thick low vegetation.

VOICE Very distinctive, explosive *chink, cher-chink* notes and tack calls.

adult

GENERAL Widespread, locally common.

Grasshopper Warbler

adult

ADULT Tiny (13 cm), skulking, dark grey-brown, heavily streaked warbler. Watch for dark-flecked crown and throat, unstreaked grey-buff underparts with pale whitish throat. Obscure grey-brown eyestripe. Sexes similar.

JUVENILE Much as adult.

HABITAT Summer visitor to much of Europe. Favours dense low shrubby vegetation.

NEST Well-concealed grass cup.

VOICE Characteristic high-pitched extended trill, often lasting for minutes, similar to an unreeling fishing line.

GENERAL Widespread, rarely numerous. Fan-tailed Warbler (*Cisticola juncidis*): smaller (10 cm), year-round resident of S Europe, reddish-brown, heavily streaked plumage, short cocked tail and plaintive **zee-eek** song flight.

Locustella luscinioides

adult

ADULT Small (15 cm), unstreaked reedbed warbler. Watch for reddish-brown upperparts and longish distinctively wedge-shaped tail. Note insectivorous beak, buff eyestripe, white underparts with buff flanks. Sexes similar.

JUVENILE Much as adult.

HABITAT Summer visitor and migrant to extensive reedbed areas across S and central Europe, less frequent in W.

NEST Cup concealed deep in reedy vegetation.

VOICE Reeling *churr* similar to Grasshopper Warbler (p.191), but lower-pitched and in shorter bursts. Ventriloquial, sings from reed stems.

GENERAL Restricted by habitat choice, but locally common in suitable areas.

Sedge Warbler

Acrocephalus schoenobaenus

ADULT Tiny (13 cm), noisy, heavily streaked warbler. Watch for boldly streaked back and unstreaked chestnut rump, pale-flecked dark crown, chestnut-buff eyestripe, short wedge-shaped tail. Inquisitive. Rarely flies far in open. Sexes similar.

JUVENILE As adult.

HABITAT Summer visitor and migrant over much of Europe except extreme S, SW and extreme N. Favours reedbeds and shrubby swamps.

NEST Well-concealed cup.

VOICE Vocal; rapid metallic repetitive jingling, twangy and chattering notes. *Tuck* alarm call.

GENERAL Widespread, often common.

adult

adult

ADULT Tiny (13 cm), slim, unstreaked brown warbler. Watch for sloping forehead, long beak, white throat and belly, buff flanks. Sexes similar.

JUVENILE Much as adult.

HABITAT Summer visitor to marshes and reedbeds across S, SW, central and W Europe.

NEST Cup suspended on several reed stems.

VOICE *Churr* of alarm. Song extended, repetitive, more musical than Sedge Warbler (p.193). Similar Marsh Warbler (*A. palustris*) of central, N and NE Europe best identified by fluid musical song with much mimicry; often in bushy habitat.

GENERAL Widespread, often locally common.

DISTRIBUTION
Summer visitor to
mainland Europe,
not Britain
Scandinavia and
Iceland

ADULT Small
(20 cm), but
thrush-sized and
among the larger
warblers. Watch
for unstreaked grey-brown upperparts, whitish underparts, bulky
build and relatively large, angular head with buff eye-stripe and
longish powerful beak. Long tail wedge-shaped at tip. Sexes similar.

JUVENILE Much as adult, buffer on underparts.

HABITAT Summer visitor or migrant to much of Europe except far
W and N. Favours extensive reedbeds.

NEST Bulky cup suspended from reeds.

VOICE Noisy; repetitive grating and metallic *gurk-gurk- gurk,
karra-karra-karra* etc. Sings from
reed stems.

adult

GENERAL Widespread, but rarely
numerous; usually heard before seen.

Hippolais icterina

ADULT Tiny (14 cm), warbler. Bright yellow breast and belly, yellow eyestripe, long beak and sloping forehead. Watch for blue legs, pale panel in closed wing, wingtips extending halfway along tail when perched. Sexes similar.

JUVENILE Much as adult.

HABITAT Summer visitor or migrant to central and N Europe, occasional elsewhere. Favours scrubby growth in woods, gardens, heath etc.

adult spring

NEST Neat grassy cup, well-concealed in bush.

VOICE Hard *tack* call; extended jingling song.

GENERAL Widespread, locally common. Replaced by Melodious Warbler *(H. polyglotta)* in S Europe: brown legs, no wing panel, closed wings only reach base of tail, gradually accelerating song.

adult female

ADULT Tiny (13 cm), very dark, very long-tailed warbler. Watch for grey back, dark red-brown breast and speckled throat. Often cocks white-edged tail. Secretive. Female slightly paler, duller and browner, pinker on breast.

JUVENILE As female.

HABITAT Year-round resident in W, SW and S Europe. Favours dense dry heath, gorse or maquis.

NEST Well-concealed cup low in vegetation.

VOICE Loud *chuck* or *churr*; brief soft scratchy warbling song.

GENERAL Locally common, but in variable numbers depending on winter weather. Subalpine Warbler *(S. cantillans)*: similar summer visitor to S Europe. Red throat and breast, white moustachial streak, brown wings and tail. Brief musical warbling song flight over heath and maquis.

Sardinian Warbler

Sylvia melanocephala

DISTRIBUTION
Resident around
Mediterranean
and N Africa

adult female

ADULT Tiny (13 cm), dark-capped warbler. Male pale grey below, darker grey above, with distinctive black hood. Female similar, but browner. In both sexes watch for characteristic white throat and red eye-ring. Skulking, but active.

JUVENILE As female, but duller and browner.

HABITAT Year-round resident in maquis and similar scrub-covered areas across extreme S Europe.

NEST Neat grass cup concealed low in vegetation.

VOICE Scolding chattering call; song a mixture of scratchy and melodious phrases, usually in bouncing song flight over scrub.

GENERAL Widespread, often locally common. As with other *Sylvia* warblers, inquisitive and can be drawn from cover by making soft squeaking noises.

ADULT Small (15 cm), but largish for a warbler. Dull grey-brown, paler below, dark grey head, blackish cheeks and white eye. Tail has white outer feathers. Sexes broadly similar.

JUVENILE As adult, paler and scaly with duller eye.

adult
female

HABITAT Summer visitor or migrant to S Europe. Favours open woodland, orchards, groves, parks.

NEST Well-concealed grass cup in bush.

VOICE Sharp *tchack* call; song in SW race repetitive, coarse and unmelodious, in SE race loud, fluting and melodious.

GENERAL Widespread, locally fairly common. Barred Warbler (*S. nisoria*) of similar habitats in north-central and E Europe: same size, drab grey-brown, white eyes, but dark cap. At close range, dark crescentic bars on underparts. Melodious song.

DISTRIBUTION
Summer visitor to Mediterranean, east to Turkey and Middle East

ADULT Tiny (13 cm), neat but dull warbler, whitish below with chestnut white throat, grey-brown above with white-edged tail. Watch for grey cap and blackish patches around eyes. Legs dark blue-grey. Sexes similar.

JUVENILE As adult, but browner.

HABITAT Summer visitor or migrant to central, W, N and NE Europe. Favours farmland with hedges and trees, woodland margins, scrubby hillsides.

NEST Neat grass cup concealed low in bush.

fresh adult

VOICE Abrupt *tack* call; distinctive song: a brief warble followed by a repetitive single-note rattle similar to Yellowhammer (p.249).

GENERAL Widespread, but rarely numerous.

Whitethroat

Sylvia communis

adult female

ADULT Tiny (14 cm), active warbler with distinctive song flight. Male has grey cap and cheeks, female brown. Watch for bright chestnut-brown wings and striking white throat. Tail brown, edged white. Legs pinkish brown.

JUVENILE As female, but duller.

HABITAT Summer visitor or migrant to much of Europe except far N. Favours heath, scrub, maquis and woodland margins or clearings.

NEST Neat grass cup concealed in vegetation near ground.

VOICE Harsh *tzchack* call; distinctive song, produced in song flight above vegetation: a rapid but cheerfully scratchy warble.

GENERAL Widespread, locally fairly common, reduced in W after droughts in African wintering grounds.

juvenile

ADULT Small (15 cm), robust warbler, almost best identified by its lack of distinctive features, but note voice and habitat. Upperparts drab olive-grey, underparts pale grey-buff. Uniformly olive-brown tail. Beak comparatively short and thick for a warbler. Legs blue. Sexes similar.

JUVENILE Much as adult.

HABITAT Summer visitor or migrant to much of Europe, not breeding in extreme S, W and N. Favours thick scrub or dense woodland undergrowth.

NEST Neat grass cup, concealed low in bush.

VOICE Abrupt *tack* call; distinctive song: an extended very melodious warble, sometimes considered second only to Nightingale (p.174).

GENERAL Widespread, occasionally fairly common.

ADULT Small (15 cm), plump, distinctive warbler. Upperparts brownish-grey, browner in female; underparts whitish tinged grey in male, buff in female. Watch for jet black cap of male, brown in female. Legs bluish.

JUVENILE As female, but with ginger-brown cap.

adult female

HABITAT Summer visitor and migrant over much of Europe, increasingly through the winter in W and S. Favours parks, gardens and woodlands with both thick undergrowth and mature, tall trees.

NEST Neat grass cup concealed low in bush.

VOICE Abrupt *tack* call; distinctive song: a melodious warble, briefer than Garden Warbler (p.202), usually ending with a phrase rising in pitch.

GENERAL Widespread, locally fairly common, scarce in winter.

DISTRIBUTION
Resident SW
Europe, Spain,
Portugal, France

adult spring

ADULT Tiny (10 cm) leaf warbler. Upperparts greenish-olive with indistinct pale eyestripe. Watch for characteristic silvery-white underparts, golden panel in wing and yellow rump conspicuous in flight. Active, usually in canopy. Legs brownish. Sexes similar.

JUVENILE Much as adult.

HABITAT Summer visitor and migrant in SW, S and south-central Europe. Favours mixed or coniferous woodland, frequently in hill country.

NEST Well-concealed grass cup, usually on or near ground.

VOICE Soft plaintive *who-eet* call; song a slow, measured trill.

GENERAL Widespread, locally fairly common.

adult
spring

ADULT Tiny (13 cm), but large among leaf warblers. Watch for bright yellow-green upperparts and canary-yellow eyestripe, throat and breast, and strikingly white belly. Legs pale pinkish. Active high in canopy. Sexes similar.

JUVENILE Much as adult.

HABITAT Summer visitor or migrant in central, W and N Europe, and to some mountain regions further S. Favours mature deciduous woodland with scanty undergrowth.

NEST Grassy cup, well concealed on or near ground.

VOICE Call **peeu** or **deeoo**; distinctive song: opens with a couple of **pee-oo** notes, accelerates into a cascading torrent of sip notes, often during song flight.

GENERAL Though widespread, only locally numerous.

Phylloscopus collybita

ADULT Tiny (10 cm) leaf warbler. Plump, brownish-olive above, whitish-buff below, dark legs. Best distinguished from Willow Warbler (p.207) by song. Sexes similar.

JUVENILE As adult, but yellower.

adult spring

HABITAT Summer visitor, migrant or year-round resident except in far N Europe. Favours woodland with mature trees.

NEST Grassy dome on or near ground.

VOICE *Hoo-eet* call; song an unmistakable series of explosive *chiff* and *chaff* notes.

GENERAL Widespread, often common.

ADULT Tiny (10 cm) leaf warbler. Yellow-olive above, yellower below, pale brown legs. Best distinguished from Chiffchaff (p.206) by song. Sexes similar.

JUVENILE As adult, but yellower.

HABITAT Summer visitor or migrant to much of Europe, breeding in far N, but not in far S. Favours woodland with dense undergrowth or scrub without trees.

adult spring

NEST Grassy dome on or near ground.

VOICE *Hoo-eet* call; song distinctive, a melodious, sparkling, descending, warbling trill ending in a flourish.

GENERAL Widespread, often common.

adult female

ADULT Tiny (9 cm): joint-smallest European bird. Plump and warbler-like, active in foliage. Olive-green back, white double wingbars in blackish wings. Watch for faint black moustachial streak, black-bordered gold crown stripe, and paler patch round large dark eye. Sexes similar (displaying male shows flame bases to crown feathers).

JUVENILE As adult, but lacking crown stripe.

HABITAT Year-round resident, migrant or winter visitor to much of Europe, summer visitor in N. Favours all woodland, also parks, gardens and farmland.

NEST Delicate mossy hammock high in tree.

VOICE Very high-pitched *tseee* call; song a series of high-pitched descending *see* notes ending in a flourish.

GENERAL Widespread, often common.

ADULT Tiny (9 cm); joint-smallest European bird. Plump and warbler-like, active in foliage. Yellow-green above, with golden-bronze shoulders. Watch for diagnostic head pattern of black bar through eye, bold white stripe between this and black-bordered fiery crest. Sexes similar.

JUVENILE Duller than adult, with faint white eyestripe.

HABITAT Year-round resident, migrant or winter visitor to S and W Europe, summer visitor to central areas. Favours all woodland, and scrub on migration.

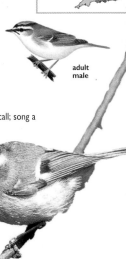

adult male

NEST Mossy hammock high in tree.

VOICE Very shrill, high-pitched **tzee** call; song a monotonous, accelerating series of **see** notes, lacking final flourish of Goldcrest (p.208).

GENERAL Widespread, locally common, scarcer in W.

juvenile

ADULT Small (15 cm), drab, rather short-legged, elongated-bodied flycatcher. Dull brown upperparts, paler underparts streaked on breast. Watch for broad, but fine, black beak. Hunts by flying out on long wings to snap up insects, often returning to same perch. Flicks wings and tail incessantly. Sexes similar.

JUVENILE As adult, but speckled on back

HABITAT Summer visitor to most of Europe. Favours woodland clearings, farmland, parks and gardens.

NEST Well-concealed shallow cup in vegetation.

VOICE Listen for distinctive **zzit** call; short squeaky song.

GENERAL Widespread, but not numerous; often inconspicuous.

ADULT Tiny (12 cm), compactly plump and distinctive flycatcher. Boldly pied summer plumage of male striking, but watch for subtler olive browns of autumn male and female. In all plumages shows broad white bar on dark wing, and white sides to base of dark tail (best seen in flight).

JUVENILE Much as female.

HABITAT Summer visitor or migrant to parts of SW, W and much of central and N Europe. Favours woodland (usually deciduous) with little undergrowth.

NEST Usually in tree hole.

VOICE Brisk *witt* call; brief, unmelodious rattling song.

GENERAL Widespread, locally fairly common.

adult
female

Aegithalos caudatus

ADULT Small (15 cm) with long, thin, black and white tail. Watch for fluffy appearance, black and white striped head pattern (all-white in far N birds), pink eye-ring, pinkish-buff shoulders and undertail. Flight feeble and whirring, calling constantly. Usually in groups. Sexes similar.

JUVENILE Much as adult, but duller and browner.

HABITAT Year-round resident in woodland, scrub, heath, farmland and gardens throughout Europe.

NEST Flask-shaped domed nest of hair, feathers and moss, camouflaged with flakes of lichen.

VOICE Noisy: thin *see-see-see* and low *tupp* calls between flock members. Rarely-heard jangling song.

GENERAL Widespread, often common.

adult northern European

ADULT Small (15 cm), tit-like, but not a true tit. Note rich chestnut-brown upperparts and long broad tail. Watch for grey head, conspicuous black 'drooping moustaches' and white throat of male. Female has brown head. Agilely clambers about reed stems. Often in groups.

JUVENILE Pale, drab version of female.

HABITAT Confined to extensive reedbeds, mostly in S and W Europe.

NEST Well-concealed cup in reeds.

VOICE Listen for frequent characteristic *ping* calls. Rarely-heard song an undistinguished rattle.

adult male summer

GENERAL Erratic in occurrence and numbers, sensitive to severe weather, but locally fairly common.

juvenile

ADULT Tiny (12 cm), black-capped tit, dull brown above, pale buffish below. Watch for neat appearance, small black bib. Best distinguished from Willow Tit (p.215) by call. Sexes similar.

JUVENILE Much as adult.

HABITAT Year-round resident over much of central and W Europe. Favours woodland, scrub and gardens.

NEST Deserted tree hole.

VOICE Explosive *pit-choo* call; song a bell-like *pitchawee-oo*.

GENERAL Widespread, but rarely numerous.

Willow Tit

Parus montanus

ADULT Tiny (12 cm), black capped tit, brownish above, pale buffish below. Watch for scruffy appearance, heavy head and neck, large dull cap and (sometimes) pale panel in wing. Best distinguished from Marsh Tit (p.214) by call. Sexes similar.

juvenile

JUVENILE Much as adult.

HABITAT Year-round resident over much of central, W and N Europe. Favours woodland, scrub and gardens.

NEST Excavates hole in rotten stump, hence thick neck muscles.

VOICE Repetitive *dee*, *chay* or *eez* notes in call; song a musical warble.

GENERAL Widespread, but rarely numerous.

ADULT Tiny (12 cm) tit with distinctive black and white chequered crest. Body brown above, buff below; watch for white face and cheeks with black 'fish-hook' marking on cheek and black bib. Sexes similar.

JUVENILE As adult, but poorly marked with little crest.

HABITAT Year-round resident over much of Europe except extreme W and SE, in mature mixed or coniferous woodland. In Scotland confined to relict ancient pine forest.

NEST Excavates tree hole.

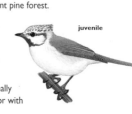

juvenile

VOICE Call characteristic purring *chirr*; song a high-pitched, repeated series of *tsee* notes.

GENERAL Widespread; occasionally fairly common. Usually solitary or with other tits.

juvenile

ADULT Tiny (12 cm) tit, active in canopy. Plump body olive or olive-grey above, pale buff below. Note white double wingbar. Watch for characteristic glossy black head with white cheeks and white patch on nape. Sexes similar.

JUVENILE As adult, duller, with grey head markings.

HABITAT Year-round resident, sometimes migrant, in European woodlands of all types, but favours mature conifers. Also gardens, parks, farmland, especially in winter.

NEST In hole or crevice in tree or bank.

VOICE High-pitched **zeet** call; song a distinctive repeated series of **wheat-zee** phrases.

GENERAL Widespread, often common. Solitary or with other tits.

Blue Tit

Parus caeruleus

ADULT Tiny (12 cm), familiar tit, active in canopy. Back green, with cobalt-blue tail and dark blue wings with a single white wingbar. Underparts yellow. Watch for distinctive head pattern of pale blue crown, white eyestripe, black stripe through eye, white cheeks and black bib. Males usually brighter than females.

JUVENILE Duller, greenish-grey instead of blue.

HABITAT Year-round resident and migrant over much of Europe except far N. Seen almost anywhere except on mountains, moors and at sea.

juvenile

NEST In hole or crevice in tree, bank or building.

VOICE *See-see-see-sit* call; song an accelerating trill after *see-see-see* notes.

GENERAL Widespread, often common, frequently gregarious.

ADULT Small (15 cm); largest European tit, often terrestrial. Note olive back, bluish wings with white wingbar. Watch for glossy black head with white cheeks, relatively long robust beak. Black bib extends in line down middle of yellow underparts; darker and more extensive in male.

JUVENILE Duller, greenish-grey instead of black.

HABITAT Year-round resident or migrant over much of Europe, summer visitor to far N. Wide habitat, but favours woodlands, parks, gardens and farmland.

NEST In hole or crevice in tree, bank or building.

juvenile

VOICE Vocal, calls very varied: *chink* most common. Song also varied, characteristic *teacher-teacher* and *see-saw* phrases.

GENERAL Widespread, often common.

220 *Sitta europaea*

adult male northern Europe

adult female

ADULT Small (15 cm), woodpecker-like. Moves head-up or head-down on branch (unlike woodpeckers). Watch for blue-grey back, white throat, black stripe through eye, longish dagger-like beak. Underparts buff, tinged deep chestnut on flanks of male. Tail short and square, white-tipped.

JUVENILE As adult, but duller.

HABITAT Year-round resident over much of Europe except far N and W. Favours deciduous woodland and parkland, occasionally gardens, especially in winter.

NEST In cavity, often in tree, usually with entrance hole plastered with mud to correct diameter.

VOICE Distinctive ringing *chwit*; whistling *too-wee*, *too-wee* song.

GENERAL Widespread, locally fairly common.

ADULT Tiny (12 cm) mouse like, creeps up tree-trunks. Mottled brown back, white underparts. Watch for bold frowning eyestripe, downcurved beak, long stiff tail, buff wingbars distinctive in undulating flight. Sexes similar.

JUVENILE As adult, but more heavily buff-speckled.

HABITAT Resident in W, central, N and NE Europe; in mature woodland favouring conifers except in Britain and Ireland.

NEST Usually in crevice behind flap of bark.

VOICE Sharp, shrill **zeee** call; song distinctive descending trill with final flourish.

GENERAL Widespread, rarely numerous. Short-toed Treecreeper *(C. brachydactyla):* almost identical, widespread in central and S Europe. Favours deciduous woods. Subtly different **zeet** call.

juvenile

Golden Oriole

ORIOLE

Oriolus oriolus

ADULT Medium (25 cm) and starling-like. Male brilliant gold and black, with shortish pink beak. Female golden olive above, whitish below with faint streaks: watch for yellow rump and yellow-tipped dark tail in flight.

JUVENILE As female, but duller, more olive.

HABITAT Summer visitor or migrant to much of Europe except N and far W. Favours mature open deciduous woodland and parkland, also orchards and groves.

NEST Grassy hammock slung between twigs.

VOICE Very characteristic fluting **wheela-wee-oo** and **too-loo-ee** calls.

GENERAL Widespread, not numerous, heard more than seen. Despite bright plumage, remarkably inconspicuous in canopy.

adult
male

ADULT Medium (35 cm), colourful crow with distinctive pinkish-buff plumage. Watch for dark-flecked crown and black moustachial streaks. Flight looks floppy and hesitant on rounded wings: look for contrasting white rump and black tail, and for blue and white wing patches. Distinctive and unusual pale pink eye. Sexes similar.

JUVENILE As adult, but duller.

HABITAT Year-round resident over much of Europe except far N. Favours woodland, parks and farmland with plentiful mature trees.

NEST Untidy twiggy shallow cup in tree fork.

VOICE Vocal; harsh *skaark* call; rarely heard soft chattering song.

GENERAL Widespread, often fairly common.

adult

Lanius collurio

adult
female

ADULT Small (18 cm) shrike. Watch for male's stubby hooked beak, rufous back, grey crown, black eyestripe and white-edged black tail. Female greyer and duller, scaly marks on breast, dark brown smudge through eye, tail brown.

JUVENILE As female, duller and heavier marks on breast.

HABITAT Summer visitor or migrant to much of Europe except far W. Favours dry open country with bushes, heath, scrub, also farmland.

NEST Neat grassy cup in bush.

VOICE Harsh *chack* call; unexpected melodious warbling song.

GENERAL Widespread, rarely numerous. Woodchat Shrike (*L. senator*), summer visitor to S Europe: white underparts, black back, chestnut and black head, bold white wingbar.

ADULT Medium (25 cm); the largest shrike. Distinctively pale and long-tailed, uses prominent perches. Watch for hooked beak, black patch through eye, white on forehead and over eye. In deeply swooping flight shows grey rump, long white-edged black tail and broad white bars in black wings. Sexes similar.

JUVENILE Similar, but browner, barred on underparts.

HABITAT Year-round resident across central Europe, winter visitor to W and S, summer visitor to N. Favours open countryside with plentiful bushes, trees and scrub.

adult female

NEST Grass cup in bush.

VOICE Harsh *chek* call; jangling song.

GENERAL Widespread, never numerous.

Nutcracker

Nucifraga caryocatactes

DISTRIBUTION
Resident of N and
E Europe from
S Sweden E to
Bering Straits, and
SC Europe

ADULT Medium
(33 cm), dull-
plumaged crow.
Watch for
straight, dark,
dagger-like beak, brown cap, white-flecked body plumage. Bold
white undertail coverts. In flight shows dark, rounded wings and
white-tipped black tail. Sexes similar.

JUVENILE Similar to adult, but duller.

HABITAT Year-round resident in conifer or
mixed forests, often mountainous, in N,
central and E Europe. Occasionally
occurs almost anywhere in W when
food short.

NEST Shallow twiggy cup in tree fork.

VOICE Vocal; harsh *skaark* and growling
calls; jangling squeaky song.

GENERAL Locally fairly common.

adult

adult

ADULT Medium (45 cm), unmistakable and familiar long-tailed pied crow. Watch for floppy flight on black and white rounded wings, long iridescent tapered tail cocked on landing. Often terrestrial. Sexes similar.

JUVENILE As adult, but duller, initially with shorter tail.

HABITAT Year-round resident over most of Europe. Wide habitat: woodland, farmland, scrub, parks and gardens.

NEST Football-size dome of twigs, high in tree.

VOICE Harsh *chack* calls and chuckles; rarely heard quiet musical warbling song.

GENERAL Widespread, locally common. Azure-winged Magpie *(Cyanopica cyanea)* of extreme SW Europe: pinkish-beige body, blue wings, long blue tail, black hood, whitish throat and collar.

ADULT Medium (37 cm), slim, glossy-black crow. Watch for shortish, yellow, slightly downcurved beak, pink legs. Often aerobatic in flight, swooping and tumbling characteristically on rounded fingered wings. Sexes similar.

JUVENILE Sooty, with grey legs and dull beak.

DISTRIBUTION
From N Spain E, through Pyrenees, Corsica, Alps, to Greece, Turkey and beyond

HABITAT Resident at high altitudes in mountains of S Europe, often near cable-car stations.

NEST Twiggy platform in cave or crevice.

VOICE Vocal; far-carrying **chee-up** or *skreee*.

GENERAL Locally fairly common. Red-billed Chough (*P. pyrrhocorax*): iridescent black with slim, downcurved crimson beak and red legs. At lower altitudes in S, on coastal cliffs in far W. **Kee-ow** call.

redbilled chough

ADULT Medium (33 cm) crow. Watch for stubby beak, black crown with slight crest, contrasting grey nape and striking white eye. In flight, has quicker wingbeats than other crows, wings rounded. Sexes similar.

JUVENILE As adult, but duller, lacking grey nape.

HABITAT Widespread resident except in N Europe. Wide habitat: woodland, farmland, city centres and coastal cliffs.

NEST Usually in tree hole, rocky cleft or building.

VOICE Metallic *jack*, also high-pitched *keeaa*.

GENERAL Widespread, often common, often gregarious. Occasionally sunbathes with seeming total relaxation, as do some other birds.

adult

adult

ADULT Medium (45 cm) crow. Watch for long, grey, dagger-like beak and bare white fleshy face contrasting with glossy iridescent black plumage. Loose feathers of upper leg give baggy-trousered appearance. Usually gregarious, often aerobatic, showing fingered wingtips. Sexes similar.

JUVENILE As adult, but duller, lacks face patch, has bristly base to straight-sided beak (see Carrion Crow, p.231).

HABITAT Year-round resident, sometimes migrant, over much of Europe, summer visitor in N, winter visitor to S. Favours farmland and open countryside with plentiful trees.

NEST Colonial, bulky twig nest high in tree.

VOICE Vocal; raucous *kaar*.

GENERAL Widespread, often common.

Carrion / Hooded Crow

ADULT Medium (45 cm) crow. All-black (Carrion) or grey with black head, wings and tail (Hooded) with intermediates where ranges overlap. Watch for black feathered base and curved ridge to beak, and neat, tight feathering to upper leg. Often solitary, in pairs or family groups, occasionally in flocks in winter. Sexes similar.

JUVENILE Much as adult, but duller.

HABITAT Carrion: year-round resident in W and SW Europe; Hooded: wider ranging through E and SE, central, N and NW, summer visitor to far N. Substantial overlap in central and S Europe. Favours all open countryside, also urban areas.

NEST Solitary bulky twig structure high in tree.

VOICE Deep harsh *korr*.

GENERAL Widespread.

adults

Raven

Corvus corax

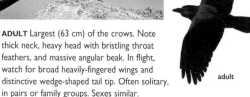

adult

ADULT Largest (63 cm) of the crows. Note thick neck, heavy head with bristling throat feathers, and massive angular beak. In flight, watch for broad heavily-fingered wings and distinctive wedge-shaped tail tip. Often solitary, in pairs or family groups. Sexes similar.

JUVENILE Similar to adult.

HABITAT Widespread year-round resident in coastal, moorland and mountain areas of W, S, E and N Europe, largely absent from central areas.

NEST Very bulky twig structure in tree, or on rocky ledge.

VOICE Gruff *pruuk* and resonant *gronk*.

GENERAL Though widespread, rarely numerous.

ADULT Small (22 cm) and familiar. Watch for iridescent black plumage with buff speckling, denser in winter. Male sings from prominent perch, throat feathers bristling, wings flapping slowly. Beak yellow in summer, black in winter. Gregarious. Fast and direct flight on triangular wings. Sexes similar.

JUVENILE Dull pale brown, darker above than below.

HABITAT Year-round resident or migrant over much of Europe, summer visitor in N, winter visitor to SW. Occurs in almost all terrestrial habitats.

NEST Untidy straw and feathers in cavity.

adult winter

VOICE Vocal; harsh shrieking calls, song full of chattering notes and mimicry of other birds.

GENERAL Widespread and common.

adult female

ADULT Small (15 cm) and familiar. Watch for typical black triangular beak and head pattern of male, with grey crown, white cheeks, brown nape and black bib. Female pale fawn on underparts, mottled browns above; note pale eyestripe.

JUVENILE Much as female.

HABITAT Year-round resident throughout Europe except extreme N. Often near habitation, favours farmland and urban areas.

NEST Untidy spherical grassy structure in dense vegetation or hole.

VOICE Harsh *chirrup*, often repetitive.

GENERAL Widespread, often common. Male Spanish Sparrow (*P. hispaniolensis*) from extreme S Europe: chestnut crown, white eyestripe, black-blotched breast; female indistinguishable from House Sparrow.

ADULT Tiny (13 cm), but chunky, sparrow. Both sexes show head pattern of brown crown and narrow white collar. Watch for white cheeks with bold black spot and small black bib.

JUVENILE As adult, but duller and browner.

juvenile

HABITAT Resident, sometimes migrant, except in far N Europe. Favours woodland, farmland and scrub.

NEST Domed grass structure, often in cavity.

VOICE Distinctively liquid **tek** and **tchup** calls.

GENERAL Widespread, sometimes fairly common. Drabber Rock Sparrow *(Petronia petronia)*: sexes similar, resembles female House Sparrow. Has indistinct yellow spot on throat, distinctive white-tipped tail. Confined to rocky areas in extreme S.

Chaffinch

Fringilla coelebs

ADULT Small (15 cm) finch. Pink breast, grey hood and black forehead in summer male, duller and masked by buff feather fringes in winter. Female olive-brown above, buff below. In flight, watch for white-edged blackish tail and bold, white, double wingbars in both sexes.

JUVENILE Much as female.

HABITAT Widespread year-round resident across Europe, summer visitor in N and E. Numbers in W and N augmented by migrants and winter visitors. Favours woodlands, farmland, parks and gardens. Often gregarious in winter.

NEST Neat, well-camouflaged cup in tree fork.

VOICE Ringing *pink* call; song a powerful cascade of rich notes ending in a flourish.

GENERAL Widespread, often common.

adult

ADULT Small (15 cm) finch. Glossy black head and back contrast with orange on breast and wings in summer male. At other times, black is partly concealed by broad orange-buff feather fringes. Female browner, but still orange on face and breast. In flight, watch for orange-white double wingbars and distinctive white rump.

JUVENILE Much as female.

HABITAT Summer visitor to N Europe, breeding in forest and woodland. Migrant or winter visitor to rest of Europe, favouring woodland, parks, gardens and farmland.

adult

NEST Neat, well-camouflaged cup in tree fork.

VOICE Drawn-out *chwaay* flight call; simple slow repetitive song based on *twee* notes.

GENERAL Widespread, but erratic.

adult
female

ADULT Tiny (10 cm) yellowish finch. Watch for streaked yellow-brown upperparts, bright yellow breast of male. Female duller and buffer. In flight, note yellow double wingbars and characteristic yellow rump contrasting with dark tail. Beak distinctively tiny and stubby, yet triangular.

JUVENILE Much as female.

HABITAT Summer visitor to W and central Europe, year-round resident further S. Favours open woodland, and farmland, parks and gardens with plenty of mature trees.

NEST Tiny neat moss and grass cup, usually high in tree.

VOICE *Churr-lit* flight call and distinctive but monotonous jingling song.

GENERAL Widespread, locally fairly common.

Greenfinch

Carduelis chloris

ADULT Small (15 cm) yellowish finch with relatively heavy, pale triangular beak. Male olive-green above, rich yellow below. Shows yellow edge to folded wing, and grey shoulders. Female duller, buffish yellow on underparts with darker streaks. In flight, fanned, slightly forked tail shows yellow patches at either side of base.

JUVENILE Much as female.

HABITAT Year-round resident or migrant over much of Europe, summer visitor to far N. Favours scrub, open woodlands, parks, gardens and farmland.

NEST Cup of twigs, moss and grass in tree or bush.

VOICE Drawn-out *dweeee* call; purring song in display flight with exaggerated wing beats.

adult

GENERAL Widespread, often common.

Goldfinch

FINCHES

Carduelis carduelis

ADULT Tiny (13 cm) colourful finch. Watch for diagnostic red face, white cheeks, black crown and nape in adult. In flight, note forked tail, white rump, black wings with distinctively broad, full-length golden wingbar. Sexes similar.

JUVENILE Dull buff, lacking head pattern, but with white rump and broad gold wingbar.

adult

HABITAT Year-round resident or migrant over much of Europe, summer visitor to parts of N and NE. Favours heath and scrub, but also open woodland, farmland, parks and gardens.

NEST Neat, well-camouflaged hair and rootlet nest, usually high in canopy.

VOICE Sharp **dee-dee-lit** call; prolonged tinkling jingle of a song.

GENERAL Widespread, often common.

adult

ADULT Tiny (12 cm), dark, agile finch. Upperparts olive, tinged yellow and heavily dark streaked. Male has black cap and bib, yellow cheeks and upper breast; female lacks black, has paler streaked breast. In flight, watch for striking yellow wingbars, yellow rump and yellow patches at base of blackish forked tail.

JUVENILE As adult female, but duller.

HABITAT Year-round resident, migrant or winter visitor to much of Europe, summer visitor in far N. Breeds in woodlands and forest, in winter favours birch and alder, sometimes on farmland or in parks and gardens.

NEST Twiggy cup high in canopy, often in conifer.

VOICE Flight call extended *chwee-ooo*; prolonged twittering song.

GENERAL Widespread, erratic, locally fairly common.

ADULT Tiny (13 cm), brownish finch. Male rich chestnut-brown on back, with white-edged dark tail, blackish wings with white panel. Watch for rich pink cap to buff head and pink breast, most conspicuous in summer. Winter male and female dull brown above, with paler streaked breast. Flight weakly fluttering, deeply undulating.

JUVENILE Much as female.

HABITAT Year-round resident or migrant over much of Europe, summer visitor in N. Favours heath, scrub and farmland, sometimes parks and gardens.

NEST Neat, well-concealed grassy cup in shrub.

VOICE Loud sweet call; twittering song.

GENERAL Widespread, locally common.

female

ADULT Tiny (13 cm) finch, aptly called the mountain linnet. Streaked dull brown above, buff with brown streaks below. Triangular beak, grey in summer, yellow in winter. Watch for single whitish wingbar and pink flush on breast and rump of summer male, sexes otherwise similar.

JUVENILE Much as female.

HABITAT Year-round resident or winter visitor to NW coastal heath, hills and moorland, summer visitor further N. May winter on coastal marshes and rough grassland.

NEST Well-concealed grassy cup low in shrub or on ground.

VOICE Nasal *chway* or *chweet*; musical jingling song.

GENERAL Local, sometimes gregarious.

adult
female

Redpoll

Carduelis flammea

ADULT Tiny (12 cm), compact, dark finch. Dark brown, heavily streaked above, paler buff below, streaked brown on breast. In summer, has dark red cap (poll), small black bib and male may have pink flush on breast. Note indistinct buff wingbar. Sexes similar in winter.

JUVENILE As winter female, lacking bib.

HABITAT Year-round resident, winter visitor or migrant to N, central and W Europe, summer visitor to far N. Favours mixed woodland, especially with birch, also scrub and open fields in winter in mixed flocks with other finches.

NEST Neat cup, usually high in tree.

VOICE *Chee-chee-chit* call; purring trill of song as circles high over trees.

GENERAL Widespread, locally common, in places increasing.

adult

ADULT Small (15–17 cm), heavily-built finch with distinctively bulky, parrot-like crossed beak. Males orange-brown or crimson-brown, females greenish-olive. Note swooping flight showing notched tail and parrot-like acrobatics, feeding on conifer cones.

JUVENILE Much as female.

HABITAT Widespread year-round resident over much of Europe, irregular visitor elsewhere, sometimes staying to breed if food supplies allow. Favours conifers, particularly spruce.

adult

NEST Flattish twiggy platform high in canopy.

VOICE Metallic *jip* or *jup* call; abrupt twittering song.

GENERAL Widespread, erratic, locally fairly common.

Bullfinch

Pyrrhula pyrrhula

ADULT Small (15 cm), thick-set finch. Male has black cap, grey back and red underparts. Female has black cap, but is suede-brown above, pinkish-fawn below. Looks heavy-headed and slow in undulating flight. Watch for white rump and purplish-black tail.

JUVENILE As female, but lacking black cap.

HABITAT Widespread year-round resident, sometimes migrant, over much of Europe except far SW and SE. Favours woodlands with dense undergrowth, scrub, farmland with hedges, occasionally parks and gardens.

NEST Fragile shallow twiggy platform in shrub.

VOICE Whistling *peeeuu*. Song a very quiet warble.

GENERAL Though widespread, rarely numerous. Usually solitary or in pairs.

adult

adult

ADULT Small (18 cm), but one of the larger finches. Large chestnut head and huge, silvery, wedge-shaped beak with grey nape, brown back and pinkish-buff underparts. In deeply undulating flight watch for white-tipped tail and broad white wingbars. Sexes broadly similar.

JUVENILE Much as adult, but browner and duller.

HABITAT Year-round resident or occasional migrant over much of Europe except far N and NW. Favours mature, usually deciduous, woodland with seeding trees.

NEST Bulky twiggy platform, high in canopy.

VOICE Explosive Robin-like *zik* call. Song a rarely-heard twittering warble.

GENERAL Widespread, rarely numerous. Secretive.

Plectrophenax nivalis

ADULT Small (17 cm) bunting. Summer male unmistakably black and white. Female and winter male streaked brown above, paler and buffer below; black-tipped yellow beak. Always shows white in closed wing and conspicuous white mid-wing triangle and white sides to tail in flight.

JUVENILE Similar to female.

HABITAT Summer visitor breeding in far N Europe, migrant or winter visitor to NW and north-central areas. Breeds on tundra and mountainsides, winters on weedy fields and marshes.

NEST Grassy cup, concealed in rocky crevice.

VOICE Plaintive *sweet* and *tew* calls; fast-moving Skylark-like song.

GENERAL Irregular, rarely numerous.

adult

adult

ADULT Small (18 cm), familiar yellowish bunting. Summer male brilliantly yellow-breasted, with yellow head showing few darker markings. Female and winter male duller and browner, but still with yellow on head and underparts. Note dark-streaked rich chestnut mantle and rump.

JUVENILE Similar to female, but duller.

HABITAT Widespread year-round resident over much of Europe, summer visitor in far N, winter visitor in extreme S. Favours heath, scrub, farmland and grassland with bushes.

NEST Grassy cup low in shrub.

VOICE Abrupt *twick* call; familiar song, a rattle of *zit* notes ending in a drawn-out wheezing *tzeeee*.

GENERAL Widespread, often fairly common.

BUNTINGS

Emberiza cirlus

adult

ADULT Small (16 cm) southern bunting. Male has grey-green hood with black and yellow markings, dark-streaked chestnut back, chestnut flanks and yellow belly. Female and winter male duller and browner, similar to female Yellowhammer (p.249). All show distinctive olive rump in flight.

JUVENILE Similar to female.

HABITAT Resident in W, SW and S Europe. Favours dry heath and scrub.

NEST Grassy cup concealed low in shrub.

VOICE High-pitched soft *tsip* call; song a monotonous rattle of *zit* notes, lacking final *tzeee* of Yellowhammer.

GENERAL Locally fairly common. Uses prominent bush-top song-posts.

DISTRIBUTION
Summer visitor to
mainland Europe

ADULT Small
(15 cm), drab
bunting. Summer
male has grey hood
and breast, with
yellow throat and eye-ring, dull chestnut underparts. Female and
winter male duller, streaked dull brown above, with buff throat and
cinnamon breast. Watch for distinctive pale eye-ring and pale
pinkish beak. Shows white-edged tail in flight.

JUVENILE Similar to female, but more uniformly buff.

HABITAT Summer visitor or migrant over much
of Europe, scarcer in W and N. Favours dry
open country and farmland with scattered
scrub, often on hillsides.

adult

NEST Well-concealed grassy cup.

VOICE Pwit, *tlip* and *chew* flight calls; slow
rasping song of several *zeeu* notes.

GENERAL Though widespread, rarely
numerous.

adult female

ADULT Small (15 cm) bunting. Summer male has striking black and white head pattern. Female and winter male browner, streaked; dark head with pale eyestripe, and blackish moustachial streaks. Tail black, white-edged.

JUVENILE Similar to female.

HABITAT Resident over much of Europe. Usually in marshy areas, occasionally elsewhere.

NEST Grassy cup concealed low in vegetation.

VOICE *Seep* and measured *see-you* calls; song a short, harsh and disjointed jangle.

GENERAL Widespread, locally common. Scarce Lapland Bunting (*Calcarius lapponicus*): chestnut nape, pale crown stripe, **ticky-tick-teeu call**; winters on coastal marshes.

ADULT Small (18 cm), but the largest bunting, and also the least distinctive in plumage. Upperparts brown, heavily darker streaked, underparts pale buff with brown streaking. Note thick-set appearance, bulky stubby beak. Shows no white in wings or tail in flight. Sexes similar, winter and summer.

JUVENILE Similar to adult.

HABITAT Erratically distributed across much of Europe except N. Favours open, dryish farmland, heath and grassland.

adult

NEST Well-concealed grassy cup on ground.

VOICE *Tsip* or quit call; song unmistakable grating metallic harsh jangle, from prominent perch.

GENERAL Though widespread, numbers very variable.